Capilano College Library

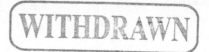

Copyright © 1995 by the Royal British Columbia Museum

All rights reserved. No part of this book may be reproduced or transmitted in any form by any means without permission in writing from the publisher, except by a reviewer, who may quote brief passages in a review.

Published by the Royal British Columbia Museum, 675 Belleville Street, Victoria, British Columbia, V8V 1X4, Canada.

Edited by Larry Eggleton.
Designed and typeset in Times 10/13 by Gerry Truscott, RBCM.
Maps and enhancements to J.R. Janish illustrations by Gerald Luxton, RBCM.
Cover illustration by Elizabeth J. Stephen.
Cover design by Chris Tyrrell.
Printed in Canada.

The publisher thanks the University of Washington Press for permission to reprint J.R. Janish's illustrations from *Vascular Plants of the Pacific Northwest*, Part 1 (L.C. Hitchcock et al, 1969).

Canadian Cataloguing in Publication Data

Pavlick, Leon E. (Leon Edward), 1939 -
   Bromus L. of North America

ISBN 0-7718-9417-1

1. Bromegrasses - North America. I. Royal British Columbia Museum. II. Title.

QK495.G74P38 1995   584'.93   C94-960291-4

Bromus L.
of North America

Leon E. Pavlick

Illustrations by Elizabeth J. Stephen, Peggy Frank and J.R. Janish

CONTENTS

Introduction   5

Key   11

Species Accounts   23

   Section Bromopsis   24

   Section Bromus   70

   Section Ceratochloa   94

   Section Genea   116

   Section Neobromus   128

Excluded Species   130

Nomenclature   131

Glossary   145

Bibliography   149

Index   157

Province of British Columbia Biodiversity Publications

The province of British Columbia has a bounty of biological diversity. Snow-clad peaks, rain-drenched forests, arid grasslands, all sizes of rivers, lakes and wetlands, and a long, rugged coast are habitats for more species of living organisms than any other province in Canada.

Because of this diversity and the size of British Columbia, there is much to be discovered about these organisms: their distribution, abundance, habitat needs and interrelationships with their environment. Increasing our knowledge about British Columbia's biodiversity will help us with the complex task of managing our land and waters sustainably.

In 1992, the provincial government initiated a co-operative biodiversity research program with funding from the Corporate Resource Inventory Initiative, the Ministry of Forests (Research Branch), the Ministry of Environment, Lands and Parks (Wildlife and Habitat Protection branches), the Ministry of Small Business, Tourism and Culture (Royal British Columbia Museum) and the Federal Resource Development Agreement (FRDA II).

One goal of this research program is to extend information to the public, managers and fellow scientists through biodiversity publications. These publications will increase awareness and understanding of biodiversity, promote the concepts and importance of conserving biodiversity and communicate provincial government initiatives related to biodiversity.

We hope these publications will be used as tools for the conservation of British Columbia's rich living legacy.

For more information, contact:
- Ministry of Forests, Research Branch, 31 Bastion Square, Victoria, B.C. V8W 3E7
- Ministry of Environment, Lands and Parks, Wildlife Branch, 780 Blanshard Street, Victoria, B.C., V8V 1X4
- Royal British Columbia Museum, 675 Belleville Street, Victoria, B.C. V8V 1X4

INTRODUCTION

This treatment of the genus *Bromus* in North America has arisen from my commitment to produce a treatment of *Bromus* for the *Manual of North American Grasses* currently being prepared (see below). My interest in this genus stems from extensive field work and research I have carried out towards preparing a new grass manual for British Columbia (in preparation) and towards solving some taxonomic problems and preparing taxonomic treatments for several complexes in the genus *Festuca* (Pavlick 1982, 1983a, 1983b, 1983c, 1984, 1985; Pavlick & Looman 1984). While doing field work for these projects, and later, for this treatment, I collected specimens of the genus *Bromus* extensively throughout British Columbia. The regions in which I collected include the Kootenay River Valley, Rocky Mountains, Columbia Mountains, Monashee Mountains, Shuswap Highlands, Thompson Plateau, Cariboo-Chilcotin, Cariboo Mountains, Peace River country, Cascade Montains, central and northern Coast Mountains, Tatshenshini country and Vancouver Island.

Towards the present treatment of *Bromus* an approximate total of 3,500 specimens were borrowed from the following herbaria: ALA, CAS, DAO, DS, L, MO, UBC, US, V, WIS and WTU (complete names and locations of these herbaria are at the end of the Introduction). These specimens, as well as those personally collected, were examined, studied and compared using standard morphological techniques. Morphological and ecological data were recorded for problematic species and complexes. Specimens were annotated to species, and in the case of some, to subspecies or variety. A list of representative specimens used in this treatment will be turned over to the *Manual of North American Grasses*; informational copies will be available by contacting Mary Barkworth, Senior Editor, Department of Biology, Utah State University, Logan, Utah, 84322-5305, U.S.A.

The taxonomic literature was used as a guide in preparing species descriptions, but for most species the final descriptions arrived at were built from my own observations. This was less true for those species for which I had scant material on hand. I gave particular attention to the descriptions of members of the sections *Bromopsis* and *Ceratochloa* as many taxonomic problems occurred in these groups and much comparing of specimens had to be done. A number of type specimens were examined and studied, especially those of some octoploid taxa of the section *Ceratochloa* (e.g. *B. breviaristatus*, *B. carinatus*, *B. luzonensis* and *B. subvelutinus*). Much time was spent on preparing keys to all sections; for the most part, the key to the section *Bromopsis* is new. Variation encountered in many species has resulted in their appearing more than once in the key. Some infraspecific taxa are keyed out in the main key; others are keyed out or otherwise treated following the species descriptions.

The taxonomic literature was extensively, and for some groups such as the complex species *B. hordeaceus*, section *Ceratochloa* and section *Bromopsis*, intensively studied. A large amount of information gathered from other authors is put together with the results of my own research. Consequently a large bibliographic section is included in this work. From this study, and from the literature, a list of taxonomic and nomenclatural synonyms was prepared and is presented. Species habitats, especially for native species, were developed from information on herbarium sheets, my own field observations and the available literature. New range maps for all species were prepared based firstly on specimens examined, and in some cases, added to from information found in the literature. Chromosome numbers are listed in the descriptions of those species for which numbers were found available.

Of 51 species of *Bromus* recognized in this work, only 3 having limited occurrence in North America, are not illustrated. Of the other 48, 30 species and 4 subtaxa are newly illustrated; 12 species and 1 variety were illustrated by Elizabeth J. Stephen, Royal British Columbia Museum, and 18 species and 3 subspecies were illustrated by Peggy Frank, freelance illustrator, Saltspring Island, B.C. The remaining 18 species were previously illustrated for Hitchcock (1969) by J.R. Janish and are here used with the kind permission of the University of Washington Press, Seattle.

I am thankful to Mary E. Barkworth, Utah State University, for encouragement and helpful information given during the preparation of this treatment, and to Jacques Cayouette, Agriculture Canada, for reviewing the manuscript, providing information and making helpful suggestions.

Although most species presented their own taxonomic problems, some groups are taxonomically difficult. One such group is the section *Ceratochloa* (which is discussed with the treatment of that section). The philosophy used in recognizing species was to use a mainly morphological-geographical approach, modified by available biosystematic evidence, and to recognize taxonomic species. For species complexes, the degree of distinctness of morphological discontinuities between taxa for recognition at the species level was judged on the merits of each case (see Davis & Heywood, 1973, p. 92). In all cases the question was asked whether a species concept for a particular taxon would be useful to those that might use this treatment for ecological, resource management or other purposes.

No new species are proposed but a number of sometimes neglected taxa are resurrected, and some taxa treated at infraspecific rank or no rank at all (in the case of mere mention in some other works) are again treated as species. In the case of the latter, returning the taxa to species level required no new combinations! Such neglected taxa, and taxa again recognized at the species rank, include *B. anomalus sens. str., B. frondosus, B. lanatipes, B. mucroglumis, B. nottowayanus, B. porteri, B. pseudolaevipes, B. pumpellianus, B. richardsonii, B. arvensis, B. commutatus, B. aleutensis, B. carinatus sens. str., B. subvelutinus, B. marginatus, B. maritimus* and *B. polyanthus*.

Description of the genus: Perennial or annual, sometimes biennial; rhizomatous in some species; culms glabrous or hairy; leaf sheaths closed i.e. with margins connate, to near their summits where the margins separate and form the throat, typically hairy, but sometimes glabrous; auricles present in some species, absent in most; ligules 0.5-6 mm long, usually erose or lacerate, hairy or glabrous; leaves mostly flat, sometimes tending to become involute; panicle usually open, wide, sometimes narrow, sometimes racemose or nearly so, sometimes dense or contracted; spikelets mostly several to many flowered, more or less terete, but laterally compressed especially in section *Ceratochloa*; rachilla disarticulating above the glumes and between the florets; glumes unequal, acute, shorter than the spikelets, the first glume typically 1-3 nerved (ranging to 7-9 nerved), the second glume typically 3-5 nerved (ranging to 9-nerved); lemmas ranging from dorsiventrally to laterally compressed and from keeled (carinate) to non-keeled, 5-13 nerved, with entire to bilobed apices, and mostly with subapical, sometimes terminal, awns; palea usually shorter than the lemma, ciliate on the keels, adnate to caryopsis; chasmogamous to facultatively cleistogamous. $2n = 14, 28, 42, 56, 70, 84$. Basic haploid chromosome number is 7, upon which several polyploid series have been built (Stebbins & Crampton, 1960). Name from *bromos* an ancient Greek name for the oat, from *broma*, food.

Bromus belongs to the subfamily *Pooideae (Festucoideae)*. Until recently it has been widely regarded as being in the Tribe *Poeae (Festuceae)*. But it has unusual rounded starch grains, a character it shares with the *Triticeae* and not with other members of *Poeae*. It is now held to belong to the Tribe *Bromeae* and is more closely related to the *Triticeae* than to the *Poeae* (Clayton & Renvoize, 1986). *Bromus* is a large genus estimated by Clayton & Renvoize (1986) to include about 150 species, with other estimates ranging between 100 and 400 species; it occurs worldwide in temperate and cool regions. Fifty-two species occur in North America north of Mexico, 28 or 29 of which are native and 23 or 24 non-native. Of the native species, 26 are perennial; 21 species (20 perennial) belong to section *Bromopsis* and 8 (6 perennial) belong to section *Ceratochloa*. The non-native species range from sporadically introduced to well established in North American landscapes, with one species only recently introduced as an agricultural grass. The non-native species consist of 20 or 21 annuals and 3 perennials and represent 4 or 5 sections: (1) *Bromopsis*, 3 species; (2) *Bromus*, 12 species; (3) *Ceratochloa*, 2 species; (4) *Genea*, 6 species; and (5) *Neobromus*, 1 species, which is variously regarded as introduced or native. Section *Bromus* is native to Eurasia, Africa and Australia. Section *Ceratochloa* is native to North America and South America, mainly in the Cordilleras.

Native perennial species of the genus *Bromus* (e.g. *B. porteri, B. pumpellianus, B. ciliatus, B. richardsonii, B. marginatus, B. polyanthus*) provide considerable forage to grazing animals in the open forests and mountainous regions of western North America. *Bromus inermis* (Smooth Brome), a rhizomatous species, was introduced into Canada from Russia in 1896; it was introduced to the Canadian northwest where it was shown to be resistant to drought, tolerant of low temperatures and useful as both a pasture and hay crop (Anstey, 1986). A number of agricultural varieties of *B. inermis* have been developed since then in both Canada and the United States and these are used across the Prairie Provinces and northern Great Plains states. "Meadow Bromegrass" is another Eurasian taxon introduced as an agricultural grass to the U.S.A. in 1966; it has been variously referred to *B. riparius* Rehm., *B. biebersteinii* R. & S., and *B. erectus* Huds. and its taxonomy remains to be worked out. Its chromosome number, $2n = 70$, and its ciliate leaf margins match that of the description given for *B. riparius* by Smith (1980). Several cultivars of Meadow Bromegrass are now used as pasture crops in Alberta, Saskachewan and the Great Plains states of the U.S.A. (see Knowles, Baron & McCartney, 1993) (Note that the common name of Meadow Brome or Meadow Bromegrass is also used for *B. commutatus* as well as for *B. erectus*). *Bromus catharticus* is grown for winter forage in the southern states of the USA.

Many of the introduced annuals are weedy species that occupy disturbed sites such as fields, waste places, road verges and overgrazed rangeland. Some are used for hay (e.g. *B. secalinus*); some form part of winter and spring forage (e.g. *B. tectorum*, unpalatable when mature). *Bromus tectorum* is also a serious weed, especially of overgrazed rangeland and winter wheat crops. (see Morrow & Stahlman, 1984; Upadhyaya et al, 1986). Some annuals of section *Genea* e.g. *B. diandrus, B. rigidus* have sharp pointed florets or fruits and long, rough awns which may injure the noses, eyes, mouth and intestines of grazing animals.

Complete names and locations of herbaria from which specimens were borrowed:
ALA   University of Alaska - Fairbanks, Alaska
CAS   California Academy of Sciences - San Fransisco, California
DAO   Biosystematics Research Institute, Agriculture Canada - Ottawa, Ontario
DS   Dudley Herbarium of Stanford University - San Fransisco, California
L   Rijksherbarium - Leiden, Netherlands
MO   Missouri Botanical Gardens Herbarium - St Louis, Missouri
UBC   University of British Columbia, Vancouver, B.C.
US   United States National Herbarium, Smithsonian Institution - Washington, D.C.
V   Royal British Columbia Museum - Victoria, B.C.
WIS   University of Wisconsin - Madison, Wisconsin
WTU   University of Washington - Seattle, Washington

KEY

1. Spikelets laterally compressed; lemmas laterally compressed and keeled at least distally.. Section *Ceratochloa*
1. Spikelets more or less terete, mostly not strongly laterally compressed; lemmas not laterally compressed, not keeled 2

2. Plants mostly perennial (except *B. texensis*) .... Section *Bromopsis*
2. Plants mostly annual 3

3. Awns geniculate and teeth of lemma aristate, except in *B. berterianus* var. *excelsus* where the awns are twisted and divaricate but not geniculate and the teeth of the lemma are acuminate................... Section *Neobromus*
3. Awns straight or divaricate, not geniculate; teeth of lemma not aristate 4

4. First glume 1 (3-) nerved; second glume 3 (5-) nerved; lemmas narrow, elongate; spikelets oblong or wedge-shaped, wider at top; awn usually longer than lemma.. Section *Genea*
4. First glume 3-5 nerved; second glume 5-9 nerved; lemmas wide, short; spikelets ovate to lanceolate; awn about as long as or shorter than the lemma
   .... Section *Bromus*

Section *BROMOPSIS*

1. Creeping rhizomes present; lemma awnless or with an awn up to 6(7.5) mm long; auricles often present; anthers 3.5-7 mm long 2
1. Creeping rhizomes absent; lemma always awned, with awns 1-12 mm long; auricles present or absent; anthers 1-7 mm long 3

2. Culm nodes usually glabrous; leaves usually glabrous; lemmas mostly glabrous or scabrous (sometimes sparsely puberulent on the margins and less frequently on the proximal part of the back); awnless or with awn up to 3 mm long; introduced and confined to disturbed sites .... (6) *B. inermis*
2. Culm nodes often pubescent; leaves often pilose; lemmas pubescent along the margins and on the keel, to entirely densely pubescent (pilose); awn usually 1-6 mm long; part of native plant communities in arctic-alpine areas, but with relict populations at lower latitudes and elevations ..... (18) *B. pumpellianus*

3. Panicle narrow, contracted, usually 2 cm or less wide, with erect, sometimes ascending branches 4
3. Panicle usually broader than 2 cm, usually open not contracted, with lower branches often spreading or patent, sometimes ascending 5

4. Blades often involute; awn 5-7 mm long; anthers 4-6.5 mm long; introduced and very sporadic .... (3) *B. erectus*
4. Blades flat; awn 2-5 mm long; anthers 2-3.5 mm long; native to mountains of Cordillera .... (20) *B. suksdorfii*

5. Culms with 9-20 nodes; sheaths overlapping; prominent flanges at collar; auricles present and prominent .... (10) *B. latiglumis*
5. Culms mostly with less than 9 nodes: sheaths overlapping or not; mostly without prominent flanges at collar; auricles absent, or if present, short and not prominent. 6

6. First glume sometimes 1-nerved 7
6. First glume sometimes 3-nerved 21

7. Second glume 3-nerved 8
7. Second glume 5-nerved 18

8. Annual; lemmas glabrous; rare endemic of Texas. .... (21) *B. texensis*
8. Perennial; lemma usually pubescent (sometimes glabrous or scabrous in forms of *B. pubescens* and *B. vulgaris*); widespread 9

9. Ligules 2-6 mm long; lemmas coarsely hairy (sometimes glabrous) on the margins; awns 6-12 mm long; plants of shaded depressions or areas of high rainfall. .... (22) *B. vulgaris*
9. Ligules 0.4-4 mm long; lemmas variously hairy; awns 2-7(-8) mm long; plants of various habitats 10

10. Culm sheaths lanate (glabrous in f. *glaber*); blades glabrous on both surfaces; lemmas usually pubescent over the back and with truncate tips; plants of southern Rocky Mountains .... (9) *B. lanatipes*
10. Culm sheaths variously hairy or glabrous, but not lanate, except in throat area in some *B. mucroglumis*; blades pilose or glabrous; lemmas variously pubescent; plants widespread 11

11. Blades densely pilose on upper or both surfaces; panicle large, open, nodding, 14-27 mm long, often with patent, drooping branches; first glume mostly 3-nerved; endemic to mountains of California .... (5) *B. grandis*
11. Plants not with the above combination of characters 12

12. Panicles erect, contracted or open, with ascending, spreading or divaricate branches; first glume 1- or 3-nerved; awns 4-7 mm long; anthers 3.5-7 mm long; mountain slopes from Washington to California .... (13) *B. orcuttianus*
12. Panicles nodding, open, with various branching; first glume mostly 1-nerved; awns various; widespread 13

13. Culms often stout; ligules 2-4 mm long; blades 6-16 mm wide; awns 3.5-7 mm long; coast line from southeastern Alaska to Oregon .... (14) *B. pacificus*
13. Culms not stout; ligules 2 mm or less long; blades narrower, various; awns various; habitat various 14

14. Sheaths retrorsely pilose with collar area hirsute, villous or glabrous; lemmas with hirsute margins and pubescent backs; awns 4-8 mm long; plants of shaded, moist woods in eastern North America .... (17) *B. pubescens*
14. Plants not with the above combination of characters 15

15. Sheaths often tufted-pilose or lanate near summit; second glume often mucronate; lower lemmas of a spikelet, in addition to having hairy margins, glabrous across the back while the upper ones have scarce, short, appressed hairs on the back, or pilose or pubescent across the back and on the margins, in which case the lemmas are 10-12 mm long; Cordilleran 16
15. Sheaths not tufted-pilose or lanate at summit; second glume not mucronate; lemmas, in addition to having hairy margins, either all glabrous or scabrous across the back, or pubescent across the back, in which case, the lemmas are

7-10 mm long; widespread across North America or of s. Texas and Mexico 17

16. Culm nodes mostly all glabrous; sheaths often pilose-tufted in the auricle position; lower lemmas of a spikelet in addition to having hairy margins, glabrous across the back while the upper ones have scarce, short, appressed hairs on the back; southern Alaska to west Texas and California .... (19) *B. richardsonii*
16. Culm nodes hairy; sheaths lanate to pilose at throat; lemmas pubescent or pilose across the back and on the margins; s. Arizona, New Mexico and Mexico .... (11) *B. mucroglumis*

17. Lemmas glabrous or scabrous over the back; anthers 2 mm or less long; widespread in moist areas of northern and Cordilleran North America, from east to west coast .... (2) *B. ciliatus*
17. Lemmas pubescent across the back; anthers 2-4 mm long; plants of west Texas and Mexico .... (1) *B. anomalus*

18. Second glume consistently 5-nerved; ligule 1 mm or less long; lower sheaths often sericeous; collar areas of leaves usually densely pilose; blades often shiny yellow-green; damp woods, ravines in eastern North America .... (12) *B. nottowayanus*
18. Second glume mostly 3-nerved, less frequently or rarely 5-nerved; ligules 0.5-3 mm long; lower sheaths not sericeous; collar areas of sheaths hirsute, villous or glabrous (densely pilose in *B. grandis*); blades not shiny yellow-green; habitats various 19

19. Panicle branches weak, often curved and drooping; lemma 8-10(-12) mm long; anthers 2-4(-5) mm long; moist woods in eastern North America .... (17) *B. pubescens*
19. Panicle branches not weak and curved (but drooping in *B. grandis*); lemmas 9.5-15 mm long; anthers 3.5-7 mm long; dry woods or open slopes in California... 20

20. Ligules 2-3 (-4) mm long; blades 15-30 cm long, the tips not prow-shaped; panicle nodding, 14-27 cm long, often with patent, drooping branches; anthers 3.5-6 mm long .... (5) *B. grandis*
20. Ligules 1-2 mm long; blades 10-16 cm long, often with prow-shaped tips; panicle erect, with ascending-spreading branches, 11-17 cm long; anthers 5-7 mm long............................. (13) *B. orcuttianus* var. *hallii*

21. Second glumes mostly or consistently 3-nerved 22
21. Second glumes sometimes or mostly 5-nerved 31

22. Ligules 2-6 mm long; first glumes mostly 1-nerved; lemmas coarsely hairy or glabrous on the margins; awns 6-12 mm long; plants of shaded depressions or areas of high rainfall in Cordillera and along Pacific coast ... (22) *B. vulgaris*
22. Ligules 3 mm long or less; awn 7 mm long or less; plants not with the above combination of characters 23

23. Blades densely pilose on upper or both surfaces; ligule 2-3 mm long; panicle large, open, nodding, 14-27 cm long, often with patent, drooping branches; first glume consistently 3-nerved; plants of montane, chaparral or coastal sage scrub zones; endemic to California .... (5) *B. grandis*
23. Plants not with the above combination of characters 24

24. Culm sheaths mostly glabrous; culm blades glabrous and often glaucous; glumes mostly glabrous; first glumes consistently 3-nerved; second glume often mucronate; lemmas pubescent across back (rarely glabrous) and with longer marginal pubescence; plants of Arizona, New Mexico and Mexico
 .... (4) *B. frondosus*
24. Plants not with the above combination of characters 25

25. Anthers 3.5-7 mm long; awns 4-7 mm long; culm nodes 2-3(-4); leaves usually with prow-shaped tips; panicle erect, open or somewhat contracted with ascending, spreading or divaricate branches; first glume 1 or 3- nerved; mountain slopes from Washington to California .... (13) *B. orcuttianus*
25. Anthers 1-4 mm long; awns 1-5 mm long; plants not with the above combination of characters 26

26. Lemmas densely hairy along margins and glabrous or scarcely pubescent on the back; first glumes mostly 1-nerved; anthers 2 mm or less long; northern or Cordilleran North America 27
26. Lemmas usually densely pubescent both on margins and across the back (sometimes nearly glabrous in *B. lanatipes*); first glume 1 or 3-nerved; anthers (1.5-) 2-4 mm long. 28

27. Culm nodes mostly all pubescent; sheaths not pilose-tufted at auricle position; blades pilose on upper or both surfaces; lemmas glabrous or scrabrous over back; moist areas; widespread in northern and Cordilleran North America
 .... *B. ciliatus*

27. Culm nodes mostly all glabrous; sheaths often pilose-tufted in auricle position; blades glabrous; uppermost lemmas with often scarce, short appressed hairs on back; montane and subalpine zones; Cordillera. . . . . (19) *B. richardsonii*

28. Culms often stout; ligules 2-4 mm long; blades 6-16 mm wide; first glumes mostly 1-nerved; awns 3.5-7 mm long; coastline from southeast Alaska to Oregon. . . . . (14) *B. pacificus*
28. Plants not with the above combination of characters 29

29. Sheaths lanate (glabrous in f. *glaber*); blades glabrous on both surfaces; first glumes mostly 1-nerved; plants of southern Rocky Mountains . . . *B. lanatipes*
29. Sheaths glabrous, pubescent or pilose, not lanate; blades pilose or glabrous; first glumes consistently or frequently 3-nerved; plants of plains and mountains of western North America 30

30. Frequently with auricles on lower leaves; midrib of culm leaves abruptly tapered just below collar; first glume with 1 or 3 nerves; plants of western Texas and Mexico . . . . (1) *B. anomalus*
30. Auricles absent; midrib of culm leaves not abruptly tapered just below collar; first glume consistently 3-nerved; widespread on plains and mountains of western North America. . . . . (15) *B. porteri*

31. Second glumes rarely 5-nerved, mostly 3-nerved; sheaths and blades usually densely hairy; above 1,500 m elevation in California 32
31. Second glumes mostly or consistently 5-nerved; plants not with the above combination of characters 33

32. Nodes 2-3; blades pubescent, 10-16 cm long, often with prow-shaped tips; panicle erect or nodding, 10-17 cm long, with ascending-spreading branches; plants of Yellow Pine to subalpine forests . . . . . (13) *B. orcuttianus* var. *hallii*
32. Nodes 3-6; blades densely pilose on upper or both surfaces, 15-30 cm long, with tips not prow-shaped; panicle large, open, nodding, 14-27 cm long, often with patent, drooping branches; plants of montane, chaparall or coastal sage scrub zones . . . . (5) *B. grandis*

33. First glumes mostly 1-nerved, rarely 3-nerved; ligule 1 mm or less long; lower sheaths often sericeous; collar areas of leaves often densely pilose; blades often shiney yellow-green; damp woods, ravines in eastern North America
. . . . (12) *B. nottowayanus*
33. First glumes 3-nerved; plants not with the above combination of characters
. . . . 34

34. Ligules of culm leaves 2-4.2 mm long; blades glabrous; glumes glabrous; lemmas densely hairy on margins and sparsely hairy or scabrous over back; Cordillera, from southern Washington to California .... (8) *B. laevipes*
34. Ligules of culm leaves 1.5 mm or less long; blades pilose to glabrous; glumes pubescent (rarely glabrous); lemmas densely hairy on margins and more or less uniformly hairy over the back; plants of California or eastern North America 35

35. Lemmas 7-11 mm long; awns 1.5-3 mm long; anthers 1.5-2.5 mm long; plants of dry, often sandy, gravelly or limestone areas of eastern North America .... (7) *B. kalmii*
35. Lemmas 10-13 mm long; awns 3-5.2 mm long; anthers 3.5-5 mm long; plants of chaparral, coastal sage scrub and woodland-savanna zones; California from San Francisco southward. .... (16) *B. pseudolaevipes*

Section *BROMUS*

1. Lemmas awnless or sometimes with awns up to 0.8 mm long, very broad, inflated; spikelets ovate .... (25) *B. briziformis*
1. Lemma awns 2-13 mm long, but if shorter then lemmas not broad nor inflated; spikelets narrower 2

2. Caryopsis thick, strongly inrolled; lemma margins inrolled; lemmas not overlapping in fruit 3
2. Caryopsis thin, weakly inrolled or flat; lemma margins open, not inrolled; lemmas overlapping in fruit.. 4

3. Anthers 3-5 mm long; awn straight, usually over 6 mm long; spikelets often purplish tinged .... (24) *B. arvensis*
3. Anthers 1-2 mm long; awn straight or flexuous, usually less than 6 mm long; spikelets not purplish tinged......................... (33) *B. secalinus*

4. Caryopsis longer than the palea; lemmas 4.5-6.5 mm long with sharply angled margins... (30) *B. lepidus*
4. Caryopsis equal to or shorter than the palea; lemmas mostly over 6.5 mm long with bluntly angled or rounded margins 5

5. Panicle branches conspicuously sinuously curved; awns 10-16 mm long and straight .... (23) *B. arenarius*

5. Panicle branches not conspicuously sinuously curved; awns usually less than 10 mm long, but if that long then divaricate, patent or recurved 6

6. Awn terete or narrow and flattened at base, arising not more than 1.5 mm from tip of lemma, straight or weakly divaricate 7
6. Awn wide and flattened at base, arising more than 1.5 mm from tip of lemma, and divaricate, patent or recurved 10

7. Panicle mostly short and dense; lemmas chartaceous with prominent (raised) nerves and often concave internerves .... (27) *B. hordeaceous*
7. Panicle longer, open; lemmas coriaceous with obscure or distinct but not prominent nerves 8

8. Anthers 3-5 mm long; panicle to 30 cm long .... (24) *B. arvensis*
8. Anthers usually less than 3 mm long; panicle usually less than 16 mm long .. 9

9. Panicle narrow, the branches usually ascending; lemmas 6.5-8 mm long with rounded margins .... (31) *B. racemosus*
9. Panicle wide, the branches spreading to ascending; lemmas 8-11.5 mm long, with bluntly angled margins .... (26) *B. commutatus*

10. Panicle erect or dense, sometimes interrupted or verticillate; panicle branches and pedicels shorter than the spikelets 11
10. Panicle nodding or erect, mostly open; some panicle branches and pedicels as long or longer than the spikelets 12

11. Lemmas 2 mm or less wide; panicle obovoid or branches verticillate
    .... (32) *B. scoparius*
11. Lemmas usually much more than 2 mm wide; panicle usually ovoid, usually dense .... (27) *B. hordeaceus*

12. Panicle erect, open; spikelets equalling or longer than the pedicels
    .... (29) *B. lanceolatus*
12. Panicle nodding, open; spikelets mostly shorter than the pedicels 13

13. Panicle racemose and usually secund; spikelets 15-70 mm long
    .... (34) *B. squarrosus*
13. Panicle mostly compound with slender, flexuous, somewhat drooping branches; spikelets 20-40 mm long .... (28) *B. japonicus*

Section *CERATOCHLOA*

1. Annual or biennial 2
1. Perennial. 6

2. Awn short (0-3.5mm) or wanting; lemma nerves prominent most of length
   .... (38) *B. catharticus*
2. Awn long (5-15mm); lemma nerves obscure, distinct or prominent on distal half
   3

3. Lemma more pilose along margins than on back; second glume about as long as lowermost lemma .... (36) *B. arizonicus*
3. Lemma scabrous, hispidulous and/or variously pubescent but not more pilose along margins than on back; second glume shorter than lowermost lemma .. 4

4. Lower panicle branches wide-spreading or patent; lemmas more or less uniformly pubescent .... (37) *B. carinatus* var. *carinatus*
4. Lower panicle branches ascending-spreading; lemmas glabrous, scabrous or pubescent toward tip 5

5. Lemma nerves prominent on distal half; lemmas with 9-11 nerves
   .... (44) *B. stamineus*
5. Lemma nerves typically obscure, sometimes conspicuous distally; lemmas with 7-9 nerves .... (37) *B. carinatus* var. *hookerianus*

6. Sheaths and blades canescent; blades tending to become involute; culms conspicuously invested at base with old leaf sheaths .... (39) *B. subvelutinus*
6. Sheaths and blades not canescent; blades flat; culm bases not conspicuously invested with old leaf sheaths 7

7. Lowermost panicle branches over 10 cm long and wide-spreading to patent, bearing 1 or 2 large spikelets at their ends............... (43) *B. sitchensis*
7. Lowermost panicle branches usually shorter, close-spreading to erect, bearing spikelets on distal half 8

8. Sheaths glabrous to scabrous 9
8. Sheaths pilose generally or only at throat 10

9. Lemmas glabrous-scabrous; panicle open or somewhat contracted; in mountains .... (42) *B. polyanthus*
9. Lemmas pubescent to pilose; panicle narrow, dense; in maritime strand .... (41) *B. maritimus*

10. Lemma glabrous, scabrous, hispidulous or puberulent 11
10. Lemmas pubescent. 12

11. Panicle open with lower branches stiffly ascending; plants of coastal sands and disturbed areas inland .... (35) *B. aleutensis*
11. Panicle narrow with lower branches erect or closely ascending; in native grasslands .... (40) *B. marginatus* var. *seminudus* Shear

12. Panicle open with lower branches stiffly ascending; plants of coastal sands and disturbed areas inland .... (35) *B. aleutensis*
12. Panicle narrow with lower branches short (but 10-20 cm long in var. *latior*), usually erect or closely ascending; plants of native grasslands. . . . (40) *B. marginatus*

Section *GENEA*

1. Lemmas 20 mm or more long 2
1. Lemmas mostly less than 20 mm long. 3

2. Panicle loose with spreading or ascending branches; callus-scar nearly circular .... (45) *B. diandrus*
2. Panicle mostly dense with branches stiffly erect; callus-scar more or less elliptical .... (47) *B. rigidus*

3. Panicle nodding, open, with branches drooping; spikelets equal to or shorter than the panicle branches. 4
3. Panicle erect, dense or nearly so; spikelets mostly longer than the panicle branches. 5

4. Panicle branches bearing 1-3 spikelets; lemmas 14-20 mm long .... (49) *B. sterilis*
4. The longer panicle branches bearing 4 or more spikelets; lemmas 9-12 mm long (50) .... (50) *B. tectorum*

5. Panicle loose, oblong-ovoid, not densely contracted, with some panicle branches 10 mm or more long, and most branches visible; lemma often 3 mm or more wide .... (46) *B. madritensis*
5. Panicle dense, obovoid, the spikelets closely clustered with the panicle branches 1-10 mm long and mostly not readily visible; lemma often 2-3 mm wide
.... (48) *B. rubens*

Section *NEOBROMUS*

Represented in our area by one species, (51) *B. berterianus*.

SPECIES ACCOUNTS

Section *Bromopsis*

1. *Bromus anomalus* Rupr. ex Fourn.   **MEXICAN BROME**

Perennial. Culms 40-90 cm tall with pubescent to glabrous nodes and glabrous internodes; with or without auricles; sheaths glabrous to pilose; ligules to 1 mm long, truncate; blades glabrous to pilose; midrib of culm leaves narrowed just below collar; panicle 10-20 cm long, open, the branches ascending or spreading; spikelets 1.5-3 cm long; glumes pubescent (glabrous), the first 1 to 3-nerved, the second 3-nerved; lemmas 7-10 mm long, pubescent across the back and on the margins; awn 1-3 mm long; anthers 2-4 mm long. 2n = 14. (A part of taxon recognized by A. S. Hitchcock; see *B. porteri* and *B. lanatipes*.)

Habitat and Distribution: Rocky slopes, canyons and plains. Western Texas and Mexico.

Major References: Hitchcock & Chase (1951); Wagnon (1952); Soderstrom & Beaman (1968).

BROMUS L. OF NORTH AMERICA — 25

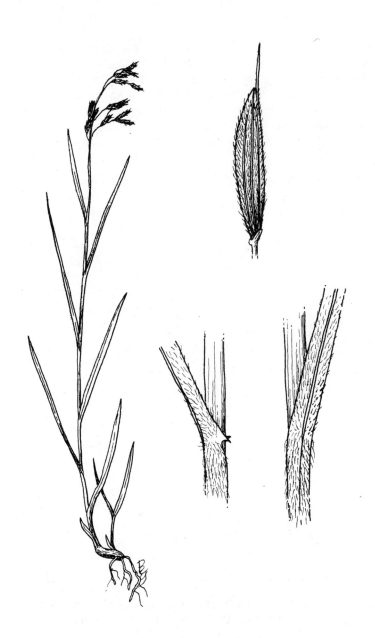

2. *Bromus ciliatus* L.   FRINGED BROME

Tufted perennial. Culms 45-120(-150) cm tall, with all nodes pubescent or lower nodes sometimes glabrous, and with glabrous internodes; without auricles; sheaths glabrous or more often the basal ones retrorsely pilose; ligules 0.4-1.4 mm long, truncate, erose, mostly glabrous (rarely pilose); blades 4-10 mm wide, 13-25 cm long, pilose on both surfaces or, more usually, glabrous at least on the lower surface; panicle nodding, 10-20 cm long, open, with ascending, spreading or drooping branches; spikelets 1.5-2.5 cm long; glumes glabrous, the first 5.5-7.5 mm long, 1(-rarely-3)-nerved, the second 6-9 mm long, 3-nerved; lemmas 9.5-14 mm long, conspicuously hirsute on the lower 1/2 to 2/3 of the margins and glabrous to scabrous over the backs; awn 3-5 mm long; anthers 1-2 mm long. 2n = 14. Specimens with densely pilose sheaths have been called forma *intonsus* (Fern.) Seymour; specimens with glabrous sheaths have been called forma *denudatus* Wiegand. (A part of the taxon recognized by A.S. Hitchcock; see *B. richardsonii*.)

Habitat and Distribution: A wide-ranging species of damp meadows, thickets, woods and streambanks; occurs at high elevations in the southern part of its range (2,000–3,000 m elevation in California). Occurs in Alaska and the Northwest Territories to Newfoundland (apparently absent from Greenland), south to Maryland, North Carolina, Illinois, Colorado and southern California (may occur in Baja California).

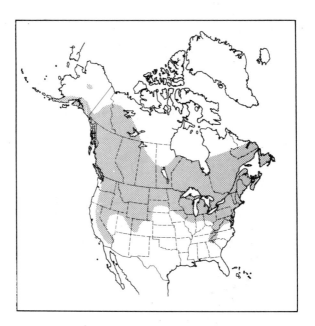

Major References: Fernald (1930); Hitchcock (1969); Hitchcock & Chase (1951); McNeill (1976); Mitchell (1965,1967); Wagnon (1952); Wiegand (1922).

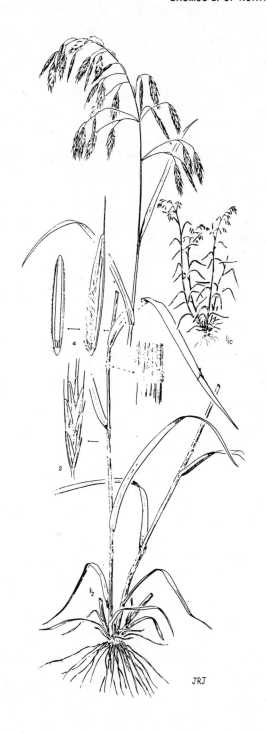

3. *Bromus erectus* Huds.   MEADOW BROME, UPRIGHT BROME

Tufted perennial. Culms 50-100 cm tall, erect with glabrous (rarely pubescent) nodes and internodes; auricles absent; sheaths glabrous to pilose; ligules to 1.5 mm, truncate; blades often involute or folded, glabrous or pilose; panicle 10-20 cm long, erect, narrow, the branches erect or ascending; spikelets 2-3 cm long; glumes glabrous, the first 1-nerved, the second 3-nerved; lemmas 10-13 mm long, glabrous or sparsely pubescent on the back and margins; awn 5-7 mm long; anthers 4-6.5 mm long. 2n = 42, 56. (See Introduction for discussion of "Meadow Bromegrass", an introduced agricultural grass that is sometimes called *B. erectus* but might belong to *B. riparius*.)

Habitat and Distribution: Disturbed ground (e.g. fields, banks, pastures, road verges) and on limestone. Europe. Introduced to North America and sporadically established in the area eastward from Wisconsin and Kentucky, and recorded from Washington, California, North Dakota and Alabama. The map does not show all areas of introduction.

Major References: Hitchcock (1969); Munz (1968); Smith (1980); Wagnon (1952);

4. *Bromus frondosus* (Shear) Woot. & Standl.

Perennial. Culms 50-100 cm tall, erect or reclining, with glabrous (rarely pubescent) nodes and glabrous internodes; auricles absent; sheaths mostly glabrous, sometimes pubescent or pilose, especially the lower ones; ligules 1-3 mm long, truncate to obtuse, laciniate, glabrous; blades 10-20 cm long, 3-6 mm wide, often glaucous, mostly glabrous or scabrous, the basal ones often pubescent; panicle 10-20 cm long, open, with branches ascending, spreading, patent or declining, drooping; spikelets 1.5-3 cm long; glumes glabrous or lightly pubescent, both 3-nerved, the first 5.5-8 mm long, the second 6.5-9 mm long, often mucronate; lemmas 8-12 mm long, often retuse, pubescent or glabrous on the back and mostly with longer marginal pubescence; awn 1.5-4 mm long; anthers 1.5-3.5 mm long. $2n = 14$.

Habitat and Distribution: Occurs in open woods, rocky slopes, often shady sites in the mountains from 1,500-2,500 meters elevation. New Mexico, Arizona, northern Mexico.

Major References: Hitchcock & Chase (1951); Shear (1900); Soderstrom & Beaman (1968); Wagnon (1952).

5. *Bromus grandis* (Shear) A. S. Hitchc.   TALL BROME

Tufted perennial. Culms 90-150 cm tall with pubescent or puberulent nodes and pubescent, puberulent to glabrous internodes; auricles short or absent; sheaths (often retrorsely) pilose (glabrous), often with collar and throat area more densely pilose, sometimes canescent on basal sheaths; ligules 2-4 mm long, obtuse, lacerate, with pilose backs; blades 5-12 mm wide, 15-30 cm long, densely pilose on both surfaces or the lower surface pubescent to glabrous; panicle nodding, 14-27 cm long, very open, with ascending to patent branches bearing spikelets near the ends and often drooping; spikelets 2-3.7 mm long; glumes pubescent, with prominent nerves, the first (1-) 3-nerved, the second 3 (-5) nerved; lemma 9.5-14 mm long, pilose on the margins and back or back pubescent to glabrous; awn 3-7 mm long; anthers 3.5-6 mm long. 2n = 14.

Habitat and Distribution: Dry wooded or open slopes in montane, chaparral or coastal sage scrub zones from about 600-2,500 m elevation. California from Placer and Tuolumne Counties southward to San Diego County, and in northern Baja California.

Major References: Hitchcock & Chase (1951); Munz & Keck (1963); Munz (1968); Shear (1900); Wagnon (1952); Wilken & Painter (1993).

6. *Bromus inermis* Leyss.   **SMOOTH BROME, HUNGARIAN BROME**

Tufted, rhizomatous perennial. Culms 50-130 cm tall, erect or spreading with glabrous (rarely pubescent) nodes and internodes; auricles short or absent; sheaths glabrous, rarely pubescent or pilose; ligules to 3 mm, truncate; blades glabrous, rarely pubescent or pilose; panicle 10-20 cm long, open, erect, the branches ascending or spreading; spikelets 2-4 cm long, sometimes near terete; glumes glabrous, the first 1 (-3) nerved, the second 3-nerved; lemmas 9-13 mm long, mostly glabrous or scabrous, sometimes sparsely puberulent on the margins and less frequently on the basal part of the back; awnless or with an awn up to 3 mm long; anthers 3.5-6 mm long. $2n = 56$.

Habitat and Distribution: Disturbed sites (e.g. fields, road verges, waste places). A major hay and forage crop in the Prairie Provinces and the northern Great Plains states. Alaska to Newfoundland, south to Mississippi, New Mexico, Arizona, California.

Major References: Anstey (1986); Armstrong (1981); Elliot (1949); Hill & Myers (1948); Mitchell (1967); Shear (1900); Wagnon (1952).

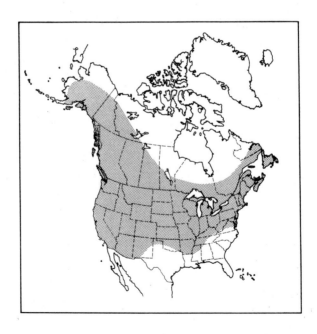

7. *Bromus kalmii* A. Gray   KALM BROME

Perennial. Culms often decumbent at the base, 50-100(-110) cm tall with pubescent (puberulent) or glabrous nodes and puberulent or glabrous internodes; without auricles; sheaths pilose to glabrous with glabrous or slightly pilose collars and pilose to glabrous throat margins; ligules 0.5-1 mm long, truncate, erose, glabrous; blades with prow-shaped tips, 5-10 mm wide, 10-17 cm long, from all glabrous to pilose on one (the upper) or both surfaces; panicle open, 8-13 cm long, drooping, few-flowered, with ascending to spreading, weak flexuous branches; spikelets 1.5-2.5 cm long; glumes pubescent, often with brown hyaline margins, the first 5-7.5 mm long, 3-nerved, the second 6.5-8.5 mm long, 5-nerved; lemmas 7-11 mm long, with densely long pilose margins and more or less uniformly pilose or pubescent backs; awn 1.5-3 mm long; anthers 1.5-2.5 mm long. 2n = 14.

Habitat and Distribution: Often sandy, gravelly or limestone soil, in openings or open woods. Minnesota through southern Ontario (but with records from near James Bay), western Quebec to New Hampshire, south to Maryland and southern Iowa (also recorded from Black Hills, South Dakota).

Major References: Dore & McNeill (1980); Gray (1848; 1859); Hitchcock & Chase (1951); McNeill (1976); Shear (1900); Wagnon (1952).

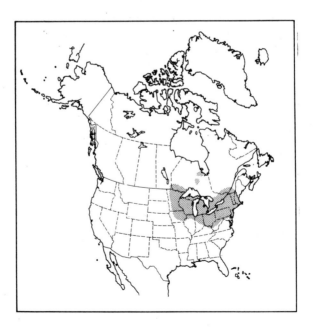

8. *Bromus laevipes* Shear

Perennial. Culms 50-150 cm tall, basally decumbent and often rooting from the lower nodes, with pubescent nodes and glabrous (rarely puberulent throughout) internodes, often puberulent-pubescent just below the nodes; auricles absent or present (short, vestigial) on basal leaves; sheaths glabrous, sometimes slightly pubescent in the summit area, sometimes with a few long hairs in the auricular position; ligules 2-4.2 mm long, obtuse, laciniate, glabrous; blades 4-10 mm wide, 13-26 cm long, glabrous (-scabrous) on both surfaces, light green or glaucous; panicle 10-20 cm long, open, nodding, with branches ascending to spreading, often drooping; spikelets 2.3-3.5 mm long, often with bronze-tinged borders on glumes and lemmas; glumes glabrous to scabrous, the first 6-9 mm long, 3-nerved, the second 8-12 mm long, 5-nerved; lemmas 12-16 mm long, densely pilose along the margins (on the proximal 1/2 or more), and sparsely pilose, pubescent or scabrous on the backs; awn 4-6 mm long; anthers 3.5-5 mm long. 2n = 14.

Habitat and Distribution: Shaded woodlands to exposed brushy slopes; between about 300-1,500 m elevation. Cordilleran; southern Washington (Klickitat Co.) southward to southern California (San Luis Obispo Co.).

Major References: Hitchcock (1969); Hitchcock & Chase (1951); Munz (1968); Munz & Keck (1963); Shear (1900); Wagnon (1952); Wilken & Painter (1993).

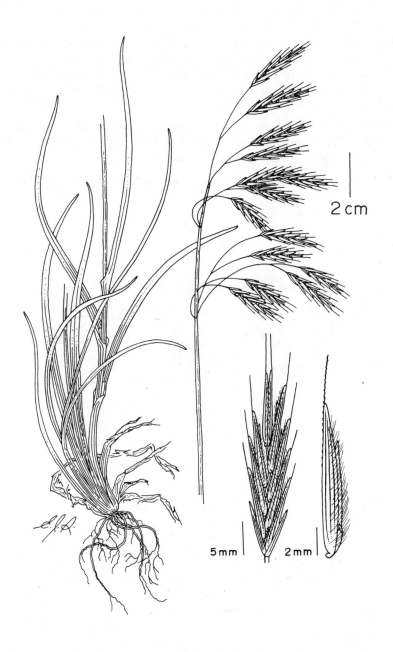

9. *Bromus lanatipes* (Shear) Rydb.   WOOLLY BROME

Perennial. Culms 40-90 cm tall, with mainly pubescent nodes and glabrous internodes (puberulent near nodes); without auricles; lower sheaths lanate and the upper ones ones nearly glabrous or all glabrous in f. *glaber*; ligules 1-2 mm long, truncate or obtuse; blades glabrous, scabrous; panicle 10-25 cm long, nodding, open, with ascending to spreading branches; spikelets 1-3 cm long; glumes glabrous (pubescent), the first 1 (-3) nerved, the second 3-nerved; lemmas 8-11 mm long, pubescent (sometimes nearly glabrous) over the back and on the margins; awn 2-4 mm long; anthers 2.5-4 mm long. 2n = 28. (*B. anomalus* var. *lanatipes* (Shear) A.S. Hitchc.)

Habitat and Distribution: Rocky or gravelly slopes, meadows, swales, brush areas, open forests in the mountains from about 800-2,500 m elevation. Colorado, Arizona, New Mexico, west Texas and Mexico.

Major References: Hitchcock & Chase (1951); Shear (1900); Wagnon (1952).

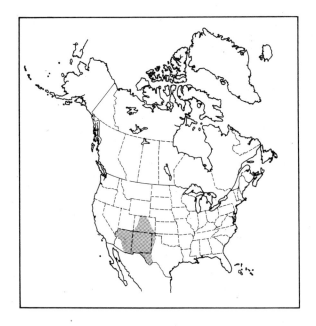

10. *Bromus latiglumis* (Shear) A. S. Hitchc. HAIRY WOODBROME
FLANGED BROME

Perennial. Culms 80-150 cm tall, with 9-20 glabrous nodes and mostly glabrous internodes (pubescent just below the nodes); auricles long, prominent; sheaths overlapping, densely or moderately retrorsely pilose to glabrous, with lanate collar areas; blades glabrous (rarely pilose) 5-15 mm wide, 20-30 cm long, with prominent flanges at their bases; ligules to 1.4 mm long, truncate, erose, hirsute and ciliate; panicle 10-22 cm long, very open, nodding, with patent, spreading or ascending branches; spikelets 1.5-3 cm long with 4-9 florets; glumes pubescent, the first 1-(rarely 3) nerved, 4-7.5 mm long, the second 3-nerved, 6-9 mm long, sometimes mucronate; lemmas 8-14 mm long, long pilose on the margins and pilose to pubescent on the back; awn 3-4.5(-7) mm long; anthers 2-3 mm long. 2n = 14. Specimens with decumbent, weak, sprawling culms, densely hairy sheaths and heavy panicles have been called f. *incanus* (Shear) Fern.

Habitat and Distribution: Shaded or open woods and thickets of stream banks, alluvial plains and slopes. Southern Saskatchewan to southern Quebec and Maine, south to Northern North Carolina, southern Missouri, Kansas and Nebraska; also reported from Edmonton, Alberta.

Major References; Hitchcock & Chase (1951); McNeill (1976, 1977); Shear (1900); Wagnon (1952).

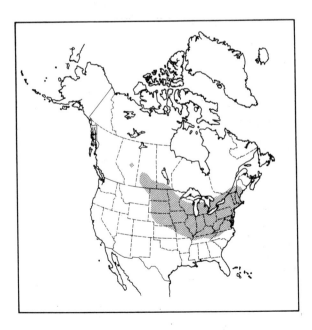

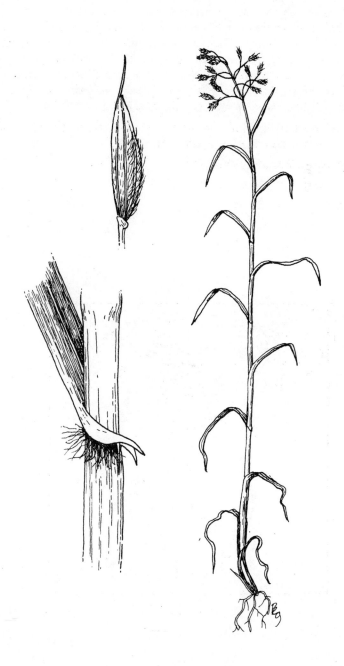

11. *Bromus mucroglumis* Wagnon

Perennial. Culms 50-100 cm tall, with pilose or pubescent nodes and glabrous internodes; auricles absent; basal sheaths pubescent to pilose with pilose or lanate throats, culm sheaths pubescent to glabrous; ligules 1-2 mm long, truncate or obtuse; blades pilose or lower surfaces glabrous; panicle 10-20 cm long, open, nodding, with branches ascending or spreading; spikelets 2-3 cm long; glumes pilose or pubescent (glabrous), the first 1-nerved, the second 3-nerved, mucronate; lemmas 10-12 mm long, pilose to pubescent on the back and margins; awn 3-5 mm long; anthers 1.5-3 mm long. 2n = 28. (not included in A.S. Hitchcock).

Habitat and Distribution: Mountains between 1,500-3,000 m elevation. Southern Arizona and southern New Mexico, northern Mexico.

Major References: Wagnon (1950a,1952)

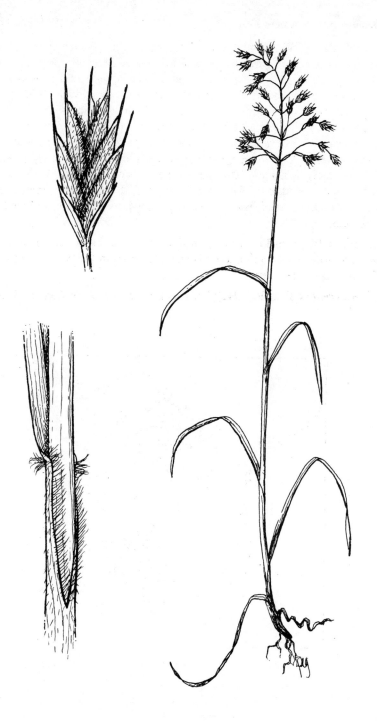

12. *Bromus nottowayanus* Fern.   VIRGINIA BROME

Perennial; Culms (60-) 70-140 cm tall with pubescent or glabrous nodes and mostly glabrous internodes; leaves without auricles; sheaths often overlapping, retrorsely pilose (glabrous), the lower ones often sericeous, with collar areas usually densely pilose; blades often shiny yellow-green, 5-12 mm wide, 15-30 cm long, with the upper surface glabrous (or pilose on the nerves) and the lower surface pilose; ligule 0.4-1 mm long, truncate, erose, ciliolate, often hairy on the back; panicle nodding, 9-25 cm long, open, with branches ascending or spreading, often recurved; spikelets 1.8-3 cm long with up to 12 florets; glumes mostly pubescent, the first 1- (rarely 3) nerved, 5.5-8 mm long, the second 5-nerved, 7-10 mm long, often mucronate; lemmas 8-13 mm long, uniformly densely pilose or sericeous or with the back more sparsely pilose or pubescent; awn 5-8 mm long; anthers 2.8-3.5 (-5) mm long. $2n = 14$.

Habitat and Distribution: Damp, shaded woods, often in ravines and along streams. Northern Iowa to southern Ontario and southern Quebec, south to northern Georgia and eastern Texas.

Major References: Fernald (1941); Hitchcock & Chase (1951); Wagnon (1952).

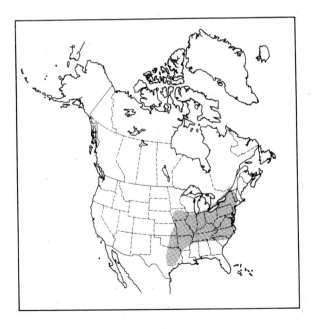

13. *Bromus orcuttianus* Vasey   ORCUTT BROME

Perennial. Culms 70-150 cm tall, erect, with pubescent (puberulent) nodes, and internodes pubescent just below the nodes or puberulent throughout; without auricles; sheaths pilose, hirsute, pubescent or glabrous; ligules 1-2 mm long, truncate or obtuse; blades usually with prow-shaped tips, glabrous or pubescent on upper or both surfaces; panicle 10-16 cm long, erect or nodding, narrow, with ascending branches or pyramidal with spreading or divaricate branches; spikelets 2-4 mm long; glumes glabrous to pubescent, the first 1-3 nerved, the second 3 (-5) nerved, sometimes mucronate; lemmas 10-16 mm long, pubescent (glabrous) on the back and pubescent to scabrous on the margins; awn 4-7 mm long; anthers 3.5-7 mm long. 2n = 14. Separated into var. *orcuttianus* and var. *hallii* Hitchc. in Jeps. as in the key. Variety *orcuttianus* has glabrous leaf blades and glabrous basal leaf sheaths, retrorsely pilose to hirsute culm leaf sheaths, and erect panicles that may be contracted with ascending branches or open with spreading or divaricate branches. Variety *hallii* has leaf blades pubescent (often densely pubescent) at least on the upper surfaces, with at least the lower leaf sheaths mostly densely pubescent, and panicles that are either erect or nodding, and open with ascending-spreading branches.

Distribution of *B. orcuttianus* var. *orcuttianus*

B. orcuttianus var. *orcuttianus*

Habitat and Distribution: Var. *orcuttianus* occurs on rocky brush slopes and in open montane forests and meadows at about 700-2,400 m elevation from southern Washington south to southern California. Var. *hallii* occurs in dry open or shaded montane to subalpine forests at about 1,100-3,000 m elevation in southern California from Monterey Co. to Riverside Co.

Major References: Hitchcock (1969); Hitchcock & Chase (1951); Jepson (1963); Munz & Keck (1963); Shear (1900); Vasey (1885); Wagnon (1952); Wilken & Painter (1993).

Distribution of *B. orcuttianus* var. *hallii*

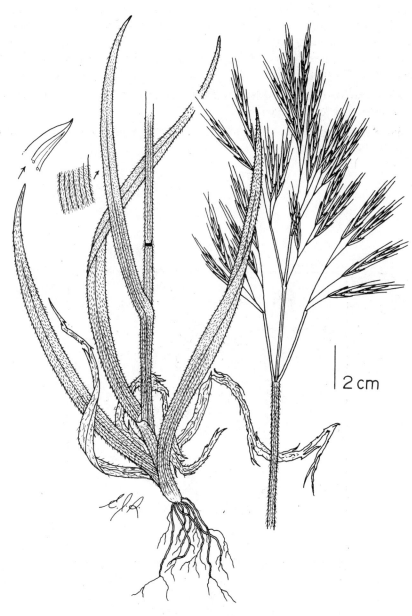

B. orcuttianus var. *hallii*

14. *Bromus pacificus* Shear   PACIFIC BROME

Perennial. Culms stout, 60-170 cm tall, erect, with pubescent nodes and glabrous (sometimes pubescent near nodes) internodes; without auricles; sheaths pilose; ligules 2-4 mm long, truncate; blades 6-16 mm wide, pilose on upper surfaces, glabrous on the lower; panicle nodding, 10-25 cm long, open with ascending, spreading or drooping branches; spikelets 2-3 cm long; glumes pubescent, the first 1 (-3) nerved, 6-8.5 mm long, the second 3-nerved, 8-11.5 mm; lemmas 10-12 mm long, pubescent on the back and more densely so on the margins; awn 3.5-7 mm long; anthers 2-4 mm long. 2n = 28.

Habitat and Distribution: Moist thickets, openings, ravines along the coast. Southeastern Alaska to Oregon.

Major References: Hitchcock (1969); Hitchcock & Chase (1951); Shear (1900); Wagnon (1952).

15. *Bromus porteri* (Coult.) Nash   NODDING BROME

Perennial. Culms 30-100 cm tall, with glabrous or pubescent nodes and glabrous (puberulent near nodes) internodes; without auricles; sheaths glabrous or pilose; ligules to 2.5 mm long, truncate or obtuse; blades glabrous or pilose on upper surface; midribs of culm leaves uniform above and below the collar; uppermost leaf often extending beyond panicle; panicle 7-20 cm long, open, nodding, often tending to be secund, with branches slender, ascending to spreading, often recurved and flexuous; spikelets 12-15 mm long; glumes pubescent (glabrous), the first 3-nerved, the second 3-nerved; lemmas 8-14 mm long, pubescent or pilose, rarely glabrous on the back and margins, often with longer hairs on the margins; awn (1-)2-3(-3.5) mm long; anthers 2-3 mm long. 2n =14. (Not recognized by A.S. Hitchcock.)

Habitat and Distribution: Mesic steppe to open forests, from about 1,000-3,200 m elevation. East of the Coast and Cascade Mountains, from central and eastern British Columbia to western Manitoba, southward to eastern California, New Mexico and west Texas.

Major References: Coulter (1885); Shear (1900); Wagnon (1952)

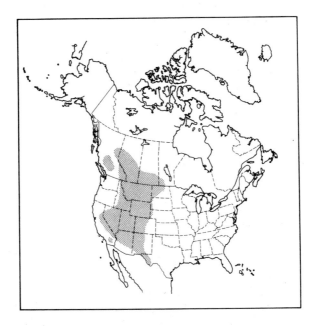

16. *Bromus pseudolaevipes* Wagnon   WOODLAND BROME

Perennial. Culms 60-120 cm tall, with pubescent or puberulent nodes and mostly glabrous (often pubescent to puberulent just below the nodes) internodes; auricles present on lower leaves, rarely absent; sheaths often pilose at the auricular position, otherwise glabrous to pilose; ligules to 1.5 mm long, obtuse, laciniate, ciliolate, pubescent on the back; blades 3-9 mm wide, 10-25 cm long, glabrous, pilose only on margins or pilose; panicle 10-20 cm long, open, mostly nodding, with branches ascending to patent to declining, often recurved; spikelets 1.5-3.5 cm long, often with glumes and lemmas bronze-tinged on borders; glumes pubescent, rarely scabrous or glabrous, the first 3-nerved, 4-7 mm long, the second (3-) 5-nerved, 6.5-9 mm long; lemmas 10-13 mm long, hirsute on the margins to near the tip, the backs mostly pubescent, sometimes becoming glabrous or scabrous distally; awn 3-5.2 mm long; anthers 3.5-5 mm long. $2n = 14$. (not recognized by A.S. Hitchcock).

Habitat and Distribution: Dry shaded or semi-shaded sites in the chaparral, coastal sage scrub and woodland-savanna zones; from near sea level to about 900 m elevation.. Found on the Coast Ranges of California from San Francisco to San Diego County, and on Santa Catalina and Santa Cruz Islands.

Major References: Munz & Keck (1963); Wagnon (1950a, 1952); Wilken & Painter (1993).

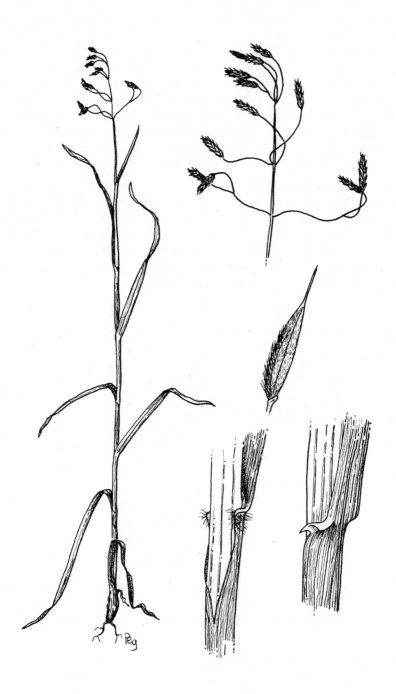

17. *Bromus pubescens* **Muhl. ex Willd.** **CANADA BROME**

Perennial. Culms 65-120 (-150) cm tall, with pubescent (glabrous) nodes and pubescent to glabrous internodes; without auricles; sheaths retrorsely pilose with collar area hirsute, villous or glabrous; ligules 0.5-2 mm long, obtuse to truncate, glabrous, erose; blades 12-32 cm long, pilose or pubescent on 1 or both sides, or all glabrous; panicle 10-25 cm long, very open, mostly nodding with weak, patent or spreading (ascending) branches that are often curved or drooping; spikelets 1.5-3 cm long; glumes pubescent (rarely glabrous), the first 4-8 mm long, 1-nerved, the second 5-10 mm long, 3(-5) nerved; lemmas 8-10 (-12) mm long, mostly with hirsute margins and pubescent backs, but sometimes all glabrous (scabrous); awn (3-)4-7 (-8) mm long; anthers 2-4 (-5) mm long. 2n = 14. (*B. purgans* L.).

Habitat and Distribution: Shaded, moist, often upland deciduous woods from North Dakota to southern Ontario and Vermont, south to western Florida and eastern Texas, and at isolated points westward (Wyoming, Nebraska). Previously reported for southern Quebec; but the only two specimens I have seen from that area identified to *B. pubescens* on their labels, I have annotated to *B. nottowayanus*.

Major References: Dore & McNeill (1980); Hitchcock & Chase (1951); McNeill (1976); Shear (1900); Wagnon (1952); Wiegand (1922).

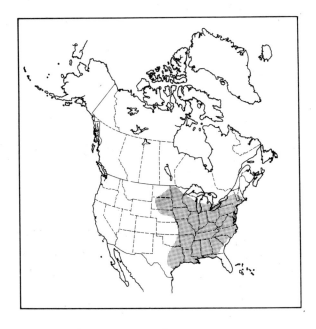

18. *Bromus pumpellianus* Scribn. PUMPELLY BROME, ARCTIC BROME

Rhizomatous perennial. Culms 50-120 cm tall, erect with puberulent, pubescent, hirsute or glabrous nodes and glabrous to pubescent internodes; short auricles present on lower leaves, or absent; sheaths pilose, villose or glabrous; ligules to 3 mm long, truncate or obtuse; blades pilose (rarely glabrous) on upper surface and glabrous or pilose on lower surface; panicle 10-20 cm long, erect or nodding, open or narrow, the branches erect to spreading; spikelets 2-3 cm long; glumes glabrous, pubescent or hirsute, the first 1-nerved, the second 3-nerved; lemmas 9-14 mm long, from pubescent along the margins and on proximal part of back to densely sericeous; awnless or (more frequently) with an awn up to 6 mm long; anthers 3.5-7 mm long. $2n = 28, 56$. Alaskan and arctic varieties separated from the typical variety on the basis of spikelet pubescence show intergradation of characters and may not be taxonomically significant – e.g., var. *pumpellianus* having glabrous glumes and lemmas especially pubescent along the margins and often on the keel or proximal part of back; var. *villosissimus* Hult. having glumes and lemmas densely long sericeous; and var. *arcticus* (Shear) Pors. with intermediate hairiness of glumes and lemmas. Two subspecies may be recognized:

1. Panicles narrow to open; plants spreading by rhizomes; culms mostly erect; nodes mostly 2-4, mostly pubescent; $2n = 56$; range of the species
   .... subsp. *pumpellianus*
1. Panicles usually open; plants tufted to short rhizomatous; culms ascending, often geniculate; nodes mostly 3-7, glabrous or pubescent; $2n = 28$; in Yukon River drainage........................ subsp. *dicksonii* Mitch. & Wilt.

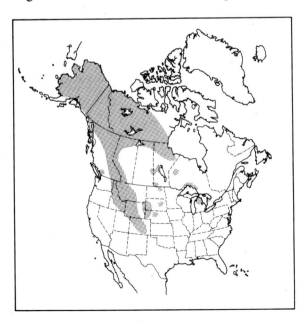

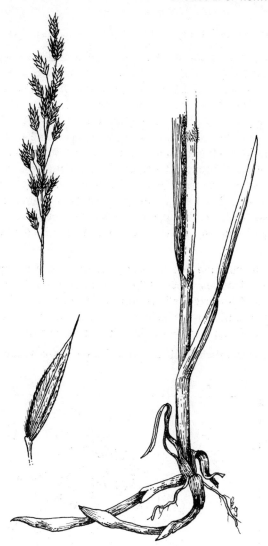

Habitat and Distribution: Sandy and gravelly streambanks and lakeshores, sand dunes, road verges, openings, meadows, dry grassy slopes (subsp. *pumpellianus*); shallow, rocky soils of riverbanks and bluffs (subsp. *dicksonii*). Found from Alaska, Yukon to northern Manitoba, southward in the Cordillera to Colorado, Black Hills, South Dakota, and at scattered points on the Great Plains of USA and Canada, the Great Lakes region and northern Ontario.

Major References: Armstrong (1981, 1983, 1984); Elliot (1949); Hitchcock & Chase (1951); Hooker (1840); Hultén (1942); Mitchell & Wilton (1966); Shear (1900); Wagnon (1952).

19. *Bromus richardsonii* Link   RICHARDSON BROME

Tufted perennial. Culms 50-110(-145) cm tall, with glabrous nodes (pubescent in some specimens from southwest U.S.A.) and internodes; without auricles; basal sheaths often retrorsely pilose, the others glabrous, often pilose-tufted at the auricle position; ligules 0.4-2 mm long, rounded, erose, ciliolate with glabrous backs; blades 3-12 mm wide, 10-35 cm long, glabrous; panicle 10-20(-25) cm long, nodding, open, with filiform branches ascending to spreading or drooping; spikelets 15-25(-40) mm long; glumes glabrous (pubescent in some specimens from southwest U.S.A.), the first 1 (rarely-3) nerved, 7.5-12.5 mm long, the second 3-nerved, 9.0-13.5 mm, often mucronate; lemmas 9-14(-16) mm long, more or less densely pilose on the lower 1/2 or 3/4 of the margins, the lower lemmas in a spikelet mostly glabrous across the back, the uppermost ones with short appressed hairs on the back; awn (2-)3-5 mm long; anthers 1-2 mm long. 2n = 28. (not recognized by A.S. Hitchcock).

Habitat and Distribution: Meadows and open woods in the upper montane and subalpine zones, between about 2,000-4,000 m elevation in the southern Rocky Mountains and at lower elevations northwards. Occurs in the Cordillera from Southern Alaska (Little Susitna Valley) and Yukon south to western Texas, Arizona and southern California (San Bernardino Mtns.) and in northern Baja California.

Major References: Mitchell (1965, 1967); Shear (1900); Wagnon (1952).

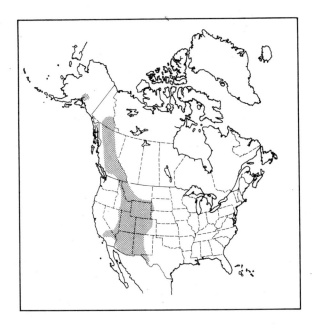

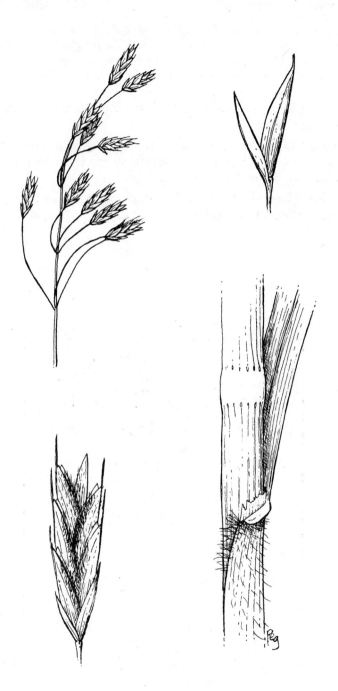

20. *Bromus suksdorfii* Vasey.   SUKSDORF'S BROME

Perennial. Culms 50-100 cm tall, erect, with glabrous nodes and internodes, or puberulent just below the nodes; auricles absent; sheaths glabrous; ligules to 1 mm long, truncate; blades glabrous with scabrous margins; panicles 6-14 cm long, erect, narrow with erect or ascending branches; spikelets 1.5-3 cm long; glumes glabrous to sparsely pubescent, the first 1 (-3) nerved, the second 3-nerved; lemmas 12-15 mm long, pubescent on the back and margins or nearly glabrous; awn 2-5 mm long. 2n = 14.

Habitat and Distribution: Open slopes, open subalpine forests, from about 1,300 -3,300 m elevation. Southern Washington (Mt. Adams) to southern Sierra Nevada of California, and western Nevada (Lake Tahoe).

Major References: Hitchcock (1969); Hitchcock & Chase (1951); Wagnon (1952); Shear (1900); Vasey (1885).

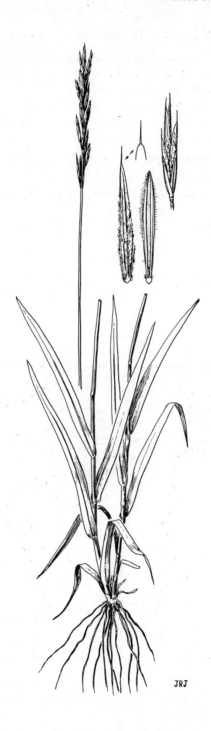

21. *Bromus texensis* (Shear) A. S. Hitchc. TEXAS BROME

Annual. Culms 30-70 mm tall, erect or spreading, with pubescent nodes; auricles absent; sheaths densely pubescent to pilose; ligules 2-3 mm long, obtuse; blades pubescent to pilose, rarely glabrous; panicle 8-15 cm long, open, drooping, few-flowered; glumes glabrous to hispidulous, the first 1-nerved, the upper 3-nerved, rarely mucronate; lemmas 9-15 mm long, smooth to scabrous; awn 4-8 mm long; anthers 3-5 mm long. 2n = 28.

Habitat and Distribution: Openings in brush areas, rocky ground. Southern Texas, northern Mexico.

Major References: Hitchcock & Chase (1951); Shear (1900); Wagnon (1952).

22. *Bromus vulgaris* (Hook.) Shear   COLUMBIA BROME

Perennial; culms 60-120 cm tall, with mostly pilose nodes and glabrous internodes; leaves without auricles; sheaths pilose to glabrous; ligules 2-6 mm long, obtuse (truncate), erose or lacerate; blades to 14 mm wide, pilose (glabrous) on upper surface, glabrous (pilose) on lower surface; panicle 10-15 cm long, open, with ascending to drooping branches; spikelets 1.5-3 cm long with 3-9 florets; glumes glabrous to pilose, the first 5-8 mm long, 1 (-3) nerved, the second 8-12 mm long, 3-nerved; lemmas 8-15 mm long, sparsely hairy to glabrous over the back and coarsely pubescent (glabrous) on the margins; awn (4-)6-12 mm long; anthers 2-4 mm long. $2n = 14$. Specimens with glabrous sheaths, usually 3-nerved first glumes, mostly 4-6 mm long awns, and which occur from southern Washington to northern California have been called var. *eximius* Shear, but are intergradient with var. *vulgaris*; robust specimens with with large panicles and pilose leaf sheaths have been described as var. *robustus* Shear.

Habitat and Distribution: Often damp, shaded or partially-shaded coniferous forests along the coast and inland in montane pine, spruce, fir, aspen forests, from sea level to about 2,000 m elevation. Coastal and central British Columbia to southwest Alberta, southward to central California and northern Utah.

Major References: Hitchcock (1969); Hitchcock & Chase (1951); Hooker (1840); Shear (1900); Wagnon (1952); Welsh et al (1987).

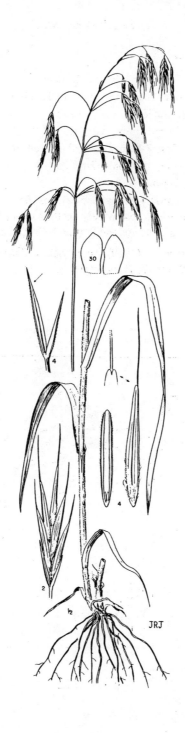

Section *Bromus*

23. *Bromus arenarius* Labill. AUSTRALIAN CHESS

Annual. Culms slender, 20-40 cm tall; sheaths densely, softly pilose; ligules 1.5-2.5 mm long, obtuse, lacerate; blades 3-6 mm wide, pilose on both sides; panicle (4-)10-15 cm long, open, nodding, with spreading or ascending sinuously curved branches; spikelets 10-20 mm long, with ca. 5-9 florets; glumes densely pilose, scarious-margined, the first 7-8 mm long, 3-nerved, the second 9-10 mm long, (5-)7 nerved; lemmas 9-11 mm long, 7-nerved, densely pilose; awn 10-16 mm long, straight; anthers 0.7-1 mm long.

Habitat and Distribution: Dry, often sandy slopes, fields, waste places. Australia. Introduced to North America; now widely scattered over California and recorded from eastern Nevada to Arizona and the Portland, Oregon area.

Major References: Hitchcock (1969); Hitchcock & Chase (1951); Jepson (1963); Munz & Keck (1963); Wilken & Painter (1993).

24. Bromus arvensis L.   FIELD CHESS

Annual. Culms 80-110 cm tall, erect, glabrous; sheaths densely, softly pilose; ligules 1-1.5 mm long, obtuse, erose, hairy; blades 10-20 cm long, coarsely pilose both sides; panicle up to 30 x 20 cm, open, erect or nodding, with long slender branches bearing up to 10 spikelets; spikelets 10-25 mm long with 4-10 florets, often purple tinged; glumes glabrous, the first, 4-6 mm long, 3-nerved, the second 5-8 mm long, 5-nerved; lemma 7-9 mm, 7-nerved, glabrous, coriaceous; awn 6-11 mm, straight; anthers 3-5 mm long, over half as long as the lemma. $2n = 14$.

Habitat and Distribution: Roadsides, fields, waste places. Southern and south-central Europe. Introduced to North America where it is of sporadic occurrence.

Major References: Hitchcock & Chase (1951); Shear (1900); Smith (1980); Wilken & Painter (1993).

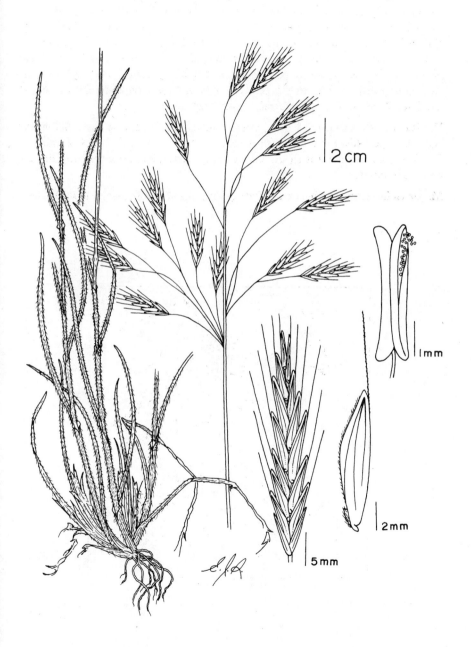

25. *Bromus briziformis* Fisch. & C. A. Mey. in Fisch., C. A. Mey. & Trautv.
RATTLESNAKE CHESS

Annual. Culms 30-62 cm tall; sheaths densely, softly pilose to lanate; ligules 0.5-2 mm long, obtuse, erose, hairy; blades pilose to pubescent on both sides ; panicle 5-15 cm long, open, secund, nodding; spikelets 15-27 mm long and ca. 10 mm wide; glumes broad, glabrous-scabridulous, the first (3-)5 nerved, 5-6 mm long, the second 7-9 nerved, 6-8 mm long; lemma 9-nerved, 9-10 mm long and over 3 mm broad, glabrous-scabridulous, coriaceous, with abruptly angled wide, hyaline margins; mostly awnless but sometimes with awns up to 0.8 mm long; anthers 0.7-1 mm long. $2n = 14$. (*B. "brizaeformis"*, orthographic error).

Habitat and Distribution: Waste places, road verges, overgrazed areas. Southwest Asia, Europe. Adventive in North America from southern British Columbia to California eastward through the Great Plains states, southern Ontario and the northeastern states.

Major References: Hitchcock & Chase (1951); Shear (1900); Smith (1980).

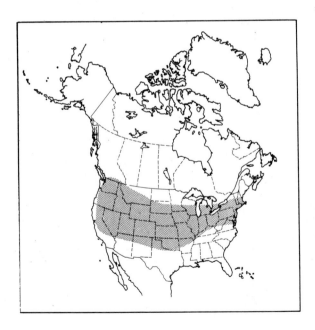

26. *Bromus commutatus* Schrad.   HAIRY CHESS, MEADOW BROME

Annual. Culms 40-120 cm tall, erect or ascending; sheaths densely, often retrorsely pilose, the upper sheaths pubescent to glabrous; ligule 1-2.5 mm long, ciliolate and glabrous or pilose on back; blades 9-18 cm long, pilose on both sides; panicle 7-16 cm long, open, with branches slender, ascending to spreading; spikelets oblong-lanceolate; glumes glabrous to scabrous (pubescent in var. *apricorum*), the first 5-7 mm long, 5-nerved, the second 6-9 mm long, 7(-9) nerved; lemma corneous, 8-11.5 mm long, distinctly (but not prominently) 7(-9) nerved, mostly glabrous on back, scabrous on margins (pubescent in var. *apricorum*), bluntly angled on the margins; awn 3-10 mm long; anthers 1-2 mm long; 2n = 14, 28, 56.

Habitat and Distribution: Fields, waste places and road verges. Southern and central Europe north to Baltic area. Introduced into North America and now found in much of northern USA and southern Canada. Resembling and perhaps conspecific with *B. racemosus*.

Major References: Amman (1981); Kerguélen (1987); Shear (1900); Smith (1980); Vivant (1964).

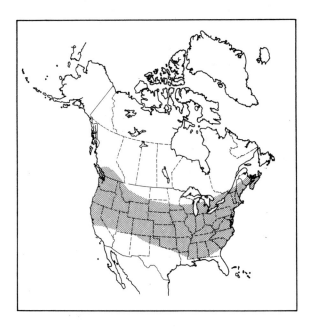

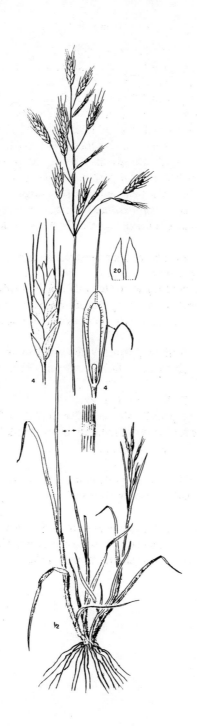

27. *Bromus hordeaceus* L. *(Bromus mollis L.)*   LOPGRASS

Annual or biennial. Culms erect or ascending, 2-50 cm tall; lower sheaths densely softly, often retrorsely pilose to lanate, the upper ones pubescent to glabrous; ligules 1-1.5 mm long, obtuse, erose, hairy; blades hairy on upper or both sides; panicle erect, mostly dense, often open at first, occasionally reduced to 1 or 2 spikelets in subsp. *thominei* and in depauperate forms of subsp. *hordeaceus*, with pedicels shorter than the spikelets; glumes pilose to glabrous, the first 5-7 mm long, 3-5 nerved, the second 6.5-8 mm long, 5-7-nerved; lemma 7-8 mm long, 7-9 nerved, chartaceous, with prominent nerves, the internerves often concave, antrorsely pilose to pubescent and proximally glabrous, to all glabrous; awns 6-8 mm long, terete or stout and flattened at base, straight or slightly curved outward or, in subsp. *divaricatus*, often twisted, divaricate or recurved at maturity; anthers 0.6-1 mm long. 2n = 28. Four taxa occur, separable as follows:

1. Awns stout, flattened at base, often twisted, divaricate, recurved or patent in fruit; panicle very dense .... subsp. *divaricatus* (Bonnier & Layens) Kerguélen (*B. molliformis* Lloyd)
1. Awns not stout, terete or narrowly flattened at base, straight or slightly curving outward in fruit; panicle mostly open at first, contracted in fruit 2

2. Culms slender, 2-16 cm tall; panicle often simple or reduced to one spikelet; awns sometimes divaricate in fruit; maritime or lacustrine sands
 .... subsp. *thominei* (Hard.) Maire in Emberger & Maire.
2. Culms usually not especially slender, often stout, (3-)10-70 cm tall; panicle mostly contracted, usually not reduced to 1 or 2 spikelets; awns straight, erect; habitat various 3

3. Lemmas 6.5-8 mm long, mostly glabrous, with broad or narrow, often abruptly angled hyaline margins; caryopsis usually as long as palea
 .... subsp. *pseudothominei* (P. Smith) H. Scholz (*B. racemosus* sensu A.S. Hitchcock, non L., in part).
3. Lemmas 8-11 mm long, mostly pilose to pubescent, with narrow, bluntly angled hyaline margins; caryopsis shorter than the palea .... subsp. *hordeaceus* (*B. mollis* L.)

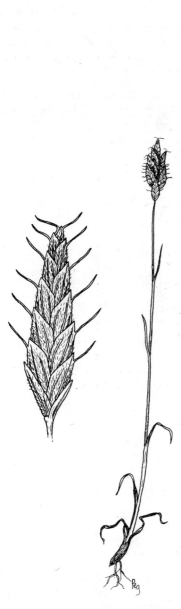
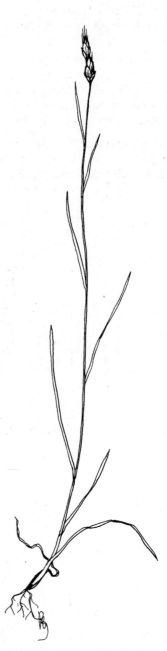

B. hodeaceus subsp. divaricatus   B. hodeaceus subsp. thominei

Habitat and Distribution: Disturbed ground, roadsides, fields, sandy beaches, waste places. Almost throughout Europe. Introduced to North America where subsp. *hordeaceus* now occurring from southern Alaska to southern Alberta, eastward through the northern Great Plains, southern Ontario to Newfoundland, south to Baja California, Missouri, North Carolina; subsp. *pseudothominei* sporadic throughout this range; subsp. *thominei* occurs along Pacific coast from Queen Charlotte Islands to Oregon, and specimens approaching it are found in Idaho and Montana; subsp. *divaricatus* occurs in California with isolated occurrences near Portland, Oregon and in Southern Michigan.

Major References: Böcher et al (1968); Holmberg (1924); Hylander (1945); Kerguélen (1981, 1987); Linnaeus (1753); Maire (1941); Maire & Weiller (1955); Scholz (1970); Seymour (1966); Smith (1968, 1980).

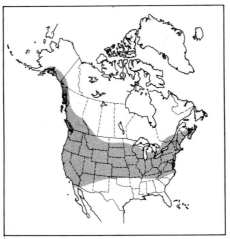

Distribution of *B. hordeaceus* subsp. *hordeaceus*

Distribution of *B. hordeaceus* subsp. *divaricatus*

B. hodeaceus subsp. *pseudothominei*   *B. hodeaceus* subsp. *hordeaceus*

28. *Bromus japonicus* Thunb. in Murr.   JAPANESE CHESS

Annual. Culms (25-)30-70 cm tall, erect or ascending; sheaths mostly very densely soft pilose to lanate, the upper ones sometimes pubescent to glabrous; ligules 1-2.2 mm long, pilose, lacerate; blades 10-20 cm long, mostly softly pilose on both sides; panicle 10-22 cm long, diffuse, nodding, with spreading to ascending, somewhat drooping, slender, flexuous branches that are mostly longer than the spikelets; spikelets 20-40 mm long with 6-12 florets; glumes glabrous to scabrous, the first 4-6 mm long, (3-)5 nerved, the second 6-8.5 mm long, 7-nerved; lemmas coriaceous, 7-9 mm long, (7-)9 nerved, smooth proximally but scabrous on distal half; awn 8-13 mm long, often divergent, often twisted, flattened at base, arising 1.5 mm or more from apex of lemma; anthers 1-1.5 mm long; 2n = 14.

Habitat and Distribution: Fields, waste places and road verges. Central and southeast Europe, Asia. Introduced into North America and now distributed throughout much of the USA and southern Canada.

Major References: Hitchcock (1969); Hitchcock & Chase (1951); Shear (1900); Smith (1980).

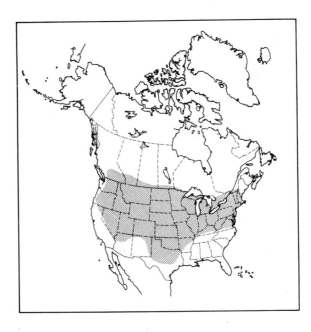

29. *Bromus lanceolatus* Roth

Annual. Culms erect or ascending, 30-70 cm tall; sheaths often densely hairy with soft white hairs; blades 10-30 cm, glabrous or pubescent; panicle erect, narrow, densely contracted when immature, 5-15 cm long, with rigid branches that are usually shorter than the spikelets; spikelets 20-50 mm long; glumes pilose, the first 5-9 mm long, the second 8-12 mm long; lemma 10-20 mm long, pilose; awns 6-12 mm long, divaricate when mature; anthers about 1 mm long. 2n = 28. (*B. macrostachys* Desf.). (Not illustrated.)

Habitat and Distribution: Waste places, or cultivated for ornament. Southern Europe. Introduced to North America and reported from few locations (e.g., Yonkers, New York (wool waste); College Station, Texas; and Washington, D.C.).

Major References: Hitchcock & Chase (1951); Smith (1980).

30. *Bromus lepidus Holmb.*

Annual, rarely biennial. Culms slender, 5-60 cm tall; panicle erect, 2-10 cm long, narrow, contracted or loose; spikelets 5-15 mm long, lanceolate, shiny; glumes glabrous; lemmas 4.5-6.5 mm long, glabrous, with sharply angled, broadly hyaline margins; awns 2-6 mm long; anthers 0.5-2 mm long; caryopsis longer than the palea. 2n = 28. (Not included in A.S. Hitchcock.) (Not illustrated.)

Habitat and Distribution: Fields, waste places. Europe. Reported from Connecticut and Massachusetts and probably occurs elsewhere in North America. Specimens of *B. hordeaceus* subsp. *pseudothominei* often approach *B. lepidus* in lemma characteristics (e.g., length, smoothness and border angle) so that either may be misinterpreted.

Major References: Holmberg (1924); Scholz (1970); Smith (1968, 1980); Seymour (1966).

31. *Bromus racemosus* L. SMOOTH BROME

Annual. Culms 20-110 cm, erect or ascending; lower sheaths densely, often retrorsely pilose, the upper sheaths glabrous to pubescent; ligules 1-2 mm long, erose, glabrous or hairy; blades 7-18 cm long, pilose on both sides; panicle 4-16 cm long, erect, narrow, the branches slender, ascending; glumes smooth to scabrous, prominently nerved, the first 4-6 mm long, (3-)5 nerved, the second 4-7 mm long, 7-nerved; lemma 6.5-8 mm long, distinctly (but not prominently) 7(-9) nerved, corneous, smooth on back, scabrous on margins, the margins rounded; awn 5-9 mm long; anthers 1.5-3 mm. $2n = 28$. (name used by A.S. Hitchcock for another taxon, *B. hordeaceus* subsp. *pseudothominei*, at least in part).

Habitat and Distribution: Fields, waste places, road verges. Western Europe north to Baltic area. Introduced into North America and now throughout much of northern USA and southern Canada. Resembles *B. commutatus* and may be a depauperate form of that species.

Major References: Ammann (1981); Kerguélen (1987); Shear (1900); Smith (1980); Vivant (1964).

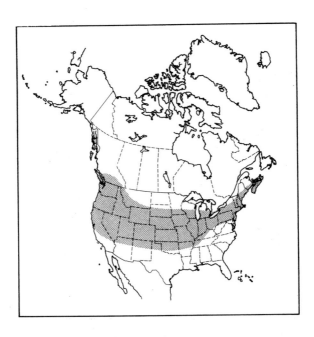

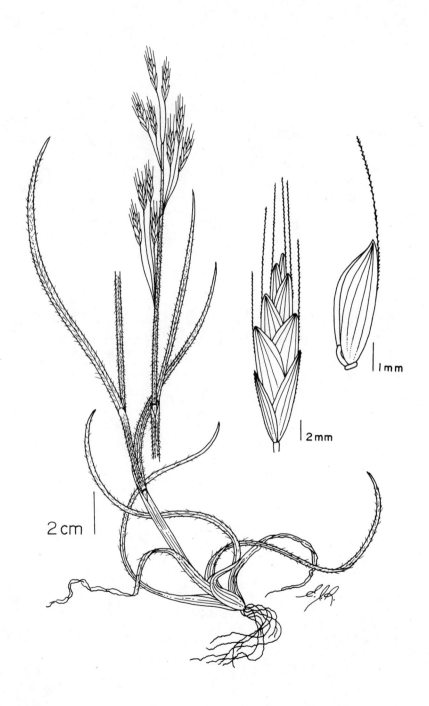

32. *Bromus scoparius* L.   BROOM CHESS

Annual. Culms 20-40 cm tall; sheaths sparsely pubescent to glabrous; blades pilose on upper surface; panicle erect, dense, 2-7 cm long, 1-3 cm wide, wedge-shaped at the base, with branches mostly shorter than the spikelets; spikelets 12-25 mm long, crowded; glumes narrow; lemmas 7-9 mm long, glabrous; awns 7-10 mm long, flattened at the base, divaricate or recurved when mature; anthers 0.3-0.5 mm long. 2n = 14. (Not illustrated.)

Habitat and Distribution: Waste places. Southern Europe. Sporadically introduced to North America; recorded from California (Mariposa), Virginia (Newport News, on ballast), and Mich. (Schoolcraft).

Major References: Hitchcock & Chase (1951); Smith (1980).

33. *Bromus secalinus* L.   RYEBROME

Annual. Culms stout (usually), erect, 20-80(-120) cm tall; sheaths glabrous to uniformly pubescent; ligules 2-3 mm long, glabrous, obtuse; blades 15-30 cm long, pilose on upper or both sides; panicle 5-23 cm long, open, nodding with branches spreading to ascending, lower branches slightly drooping, often secund after flowering; spikelets 10-20 cm long, ovoid-lanceolate or ovate, somewhat turgid, with rachilla visible and florets ascending-spreading after flowering; glumes corneous, scabrous to glabrous, distinctly or obscurely nerved, the first 3-5 nerved, 4-6 mm long, the second 7-nerved, 6-7 mm long; lemmas 6.5-8.5(-10) mm long, elliptic, obtuse, corneous, obscurely 7-nerved, mostly smooth on back and scabrous to puberulent on margins and near apex, sometimes pubescent (var. *velutinus*), with inrolled margins; awns (0-)3-6(-9.5) mm long, straight or flexuous; anthers 1-2 mm long; grain 6-9 mm long, strongly inrolled at maturity. $2n = 28$.

Habitat and Distribution: Fields, waste ground, roadsides. Introduced and now occurs more or less throughout USA and southern Canada. Southern and central Europe. Specimens with pubescent spikelets are sometimes recognized as var. *velutinus*.

Major References: Hitchcock (1969); Hitchcock & Chase (1951); Smith (1980).

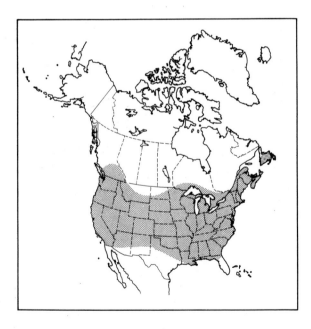

34. *Bromus squarrosus* L. ONE-WAY BROME

Annual. Culms 20-60 cm tall, erect or geniculately ascending; lower sheaths softly, densely pilose to lanate; ligules 1-1.5 mm long, obtuse, erose, ciliolate, hairy on back; blades 5-15 cm long, softly and densely pilose on both sides; panicle nodding, 7-20 cm long, open, often with few spikelets, racemose or subracemose, secund, with branches equalling or shorter than the spikelets; spikelets 15-70 mm by 7-15 mm, somewhat inflated, broadly oblong or ovate-lanceolate with 8-30 florets; glumes glabrous-scabrous, the first 4.5-7 mm long, 3-5(-7) nerved, the second 6-9 mm long, 7-nerved; lemma 8-11 mm long, 7-9 nerved, chartaceous, glabrous-scabridulous, with broad hyaline abruptly obtuse-angled borders; awn 8-10 mm long, flattened and sometimes twisted at the base, divaricate at maturity, arising 1.5 mm or more from the lemma apex ; anthers 1-1.3 mm long. 2n = 14.

Habitat and Distribution: Overgrazed pastures, fields, waste places, road verges. Europe south of central France and central Russia. Introduced to North America and now occurring from southeastern British Columbia and eastern Washington to southern Manitoba and Kansas, and sporadically in southern Ontario and the New England states.

Major References: Hitchcock & Chase (1951); Shear (1900); Smith (1980).

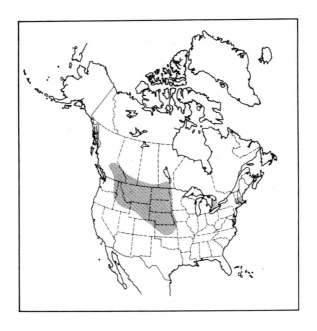

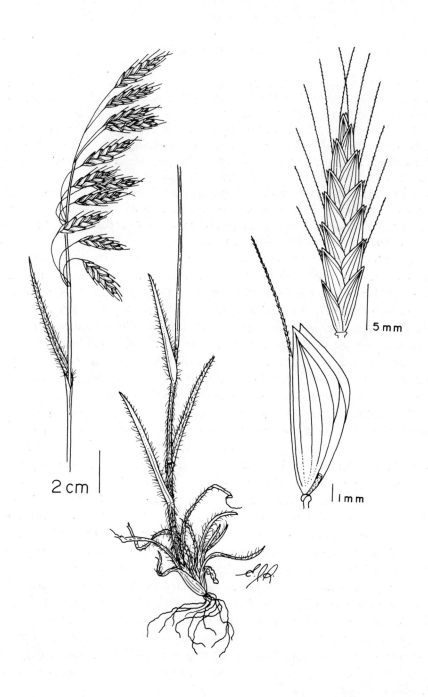

Section *Ceratochloa*

Section *Ceratochloa* contains a polyploid complex in which hybridization with members of other sections of *Bromus* is involved. This so-called *Bromus carinatus* complex is named after one of its earliest recognized species; a chromosome count of 2n = 56 is the most usual number reported for this complex, which is here taken to be represented by 7 species (*B. aleutensis*, *B. carinatus* sens. str., *B. marginatus*, *B. maritimus*, *B. polyanthus*, *B. sitchensis* and *B. subvelutinus*). These are basically species recognized in Hitchcock & Chase's *Manual of the Grasses of the United States* (1951), although numerous changes are made in concepts, descriptions and ranges. In addition to these octoploids we have in our study area one duodecaploid (2n = 84), *B. arizonicus*, and 2 hexaploids (2n = 42), *B. catharticus* and *B. stamineus*, both introduced from South America.

There is evidence that some members of the *B. carinatus* complex are interconnected by forms with which partially fertile hybrids are produced (see Harlan, 1945a, 1945b; Stebbins & Tobgy, 1944; Stebbins, Tobgy & Harlan, 1944; Stebbins, 1947; Stebbins, 1981). This and the presence of morphological intergradation of characters between some recognized species may have influenced some recent workers to lump a number of species from the North American Cordillera under the name *B. carinatus* (cf. C. L. Hitchcock, 1969; Soderstrom & Beaman, 1968; Stebbins, 1981; Welsh et al, 1987; Wilken & Painter, 1993). This *B. carinatus* *sensu lato* is a morphologically quite diverse taxonomic complex.

Cytogenetically *B. catharticus* and *B. stamineus*, both South American species, and *B. arizonicus* are distinct from the other species considered here (Stebbins & Tobgy, 1944; Stebbins, Tobgy & Harlan, 1944; Stebbins, 1981). The presence of some hybrid sterility, resulting from crosses between plants regarded as separate species on the basis of morphology, "supports the recognition of more than one species within the octoploid members of the *B. carinatus* complex, but the number of species present cannot yet be determined" (Stebbins & Tobgy, 1944). But Stebbins (1981) expressed the belief "that all the North American octoploid (2n = 56) populations should be united into a single species, in spite of the barriers of hybrid sterility that separate many of them". In order to do this one would also have to ignore some major differences in morphology between recognizable taxa and some separate ecological amplitudes and geographical ranges. And the resulting species concept would be too wide to avoid confusion when attempting to apply it to practical field situations; for example, the F_1 hybrid between *B. carinatus* sens. str. from coastal California and *B. polyanthus* from Arizona are highly sterile (Stebbins, 1981); these two species are therefore not interchangeable (e.g., in ecological restorations).

A fair question might be whether the species of the *B. carinatus* complex should not be recognized at the subspecies level. The several considerations following suggest otherwise. Despite some intermediate forms, most populations of *Ceratochloa* are often easily recognized to species on the basis of their morphology and often occupy different ecological niches and/or geographical zones. There are a number of morphological characters which separate each of the taxa. And provided some allowance is made for local hybridization, we must be able, for communication, to refer most of our specimens to binomials (Davis & Heywood, 1973). And to the ecologist and resource manager the recognition of *B. marginatus*, *B. subvelutinus*, *B. polyanthus*, *B. carinatus* sens. str., *B. sitchensis* and *B. aleutensis* can be very useful.

35. *Bromus aleutensis* Trin. ex Griseb.   ALEUT BROME

Perennial; culms mostly stout, 40-130 cm tall, often with a decumbent base, usually pubescent just below nodes; sheaths coarsely striate, sparsely retrorsely pilose to moderately pilose or pilose only at the throat; ligule 3.5-5 mm, glabrous (pubescent), laciniate; rarely with auricles; blades 6-15 mm wide and 13-35 cm long, sparsely to moderately pilose on both sides (glabrous); panicle lax, erect, 10-28 cm long with the 1 or 2 lower branches stiffly ascending and each bearing (1-)2-3 spikelets on distal half; spikelets laterally compressed, 2.2-4 cm long, with 3-6 florets; glumes glabrous to pubescent, the first 9-13 mm long, 3-5 nerved, the second 10-15 mm long, 7(-9) nerved; lemmas compressed-keeled, 12-17 mm long, with 9(-11) nerves that are conspicuous at least on distal half, soft pubescent (glabrous), firm-textured but with medium to wide hyaline margins; awn (3-)5-10 mm long; anthers 2.2-4.2 mm long; 2n = 56.

Habitat and Distribution: In sands, gravels and disturbed soil; along the coast from the Aleutian Islands to the Olympic Mountains, some lakeshores of central British Columbia (e.g., Stuart Lake) and sporadically eastward in British Columbia, southeastern Alberta, Manitoba, Ontario and Quebec, where it is possibly adventive with cultivation and road verges.

Major References: Hitchcock & Chase (1951); Hultén (1942); Shear (1900): Stebbins & Tobgy (1944).

36. *Bromus arizonicus* (Shear) Stebbins   ARIZONA BROME

Annual; culms 30-60 cm tall, erect: sheaths retrorsely pilose to glabrous with a few hairs at the throat; ligules 1-2 mm long, obtuse, erose, mostly glabrous; blades 8-18 cm long, sparsely pilose on both surfaces to glabrous on lower surface; panicle 12-25 cm long tending to be narrow and somewhat contracted with ascending branches; spikelets laterally compressed, 18-25 mm long; glumes glabrous-scabrous, subequal, the first 8-12.5 mm long, 3-nerved, the second 9.5-14 mm long, 7-nerved, about as long as the lowermost lemma; lemmas 9.5-14 mm long, distinctly 7-nerved, often pilose on the margins and pubescent over the back or on distal parts only, shortly bifid at apex; awns 6-13 mm long; anthers 0.3-5.0 mm long; 2n = 84.

Habitat and Distribution: Disturbed soils of road verges, and dry, open places, mostly below 1,000 m elevation.. In California from Yolo and Fresno Counties to San Diego Co., eastward to Arizona; also Baja California and Santa Barbara Islands.

Major References: Hitchcock & Chase (1951); Munz & Keck (1963); Shear (1900); Stebbins, Tobgy & Harlan (1944); Stebbins (1981); Wilken & Painter (1993).

37. *Bromus carinatus* Hook. & Arn.   CALIFORNIA BROME

Annual or biennial (to perennial in var. *hookerianus*); culms erect, 50-100 cm tall; sheaths retrorsely soft pilose to glabrous-scabrous; ligule 2-3(-4) mm long, obtuse, lacerate or erose, mostly glabrous; blades flat, 3-6 mm wide and 10-30 cm long, sparsely pilose on both sides (mostly) to glabrous; panicle lax, 15-40 cm long, with lower branches ascending, spreading, or, in var. *carinatus*, more often patent to declined; spikelets laterally compressed, 2-4 cm long; first glume 8-10 mm long, 3-7 nerved, the second 9.5-12 mm long, 5-9 nerved; lemma 12-16(-20) mm long, proximally obcompressed and keeled distally, with revolute margins having hyaline borders (broad hyaline borders in var. *hookerianus*) that are often upwardly recurved, mostly pubescent, varying to scabrous in var. *hookerianus*, obscurely 7-9 nerved; awn (6-)8-15 mm long; anthers 0.5-4.5 mm long. 2n = 56.

Bromus carinatus intergrades with *B. marginatus* and *B. subvelutinus*, both of which occur eastward of its range and mostly at higher elevations. Several varieties of *B. carinatus* have been described which also intergrade.

Specimens with 3-4 cm long spikelets, 20-40 cm long panicles with lower branches spreading to ascending, mostly scabrous lemmas, and broad hyaline margins, belong to var. *hookerianus* (Thurb.) Shear. Specimens from southern California having nearly smooth leaves and sheaths, and scabrous lemmas have been called var. *californicus* (Nutt.) Shear. And caespitose specimens from California having short, narrow, linear leaves and narrow subracemose panicles have been called var. *linearis* Shear; this may be intermediate between var. *carinatus* and *B. subvelutinus*.

Bromus carinatus, excluding range east of Cascade – Sierra Nevada summits.

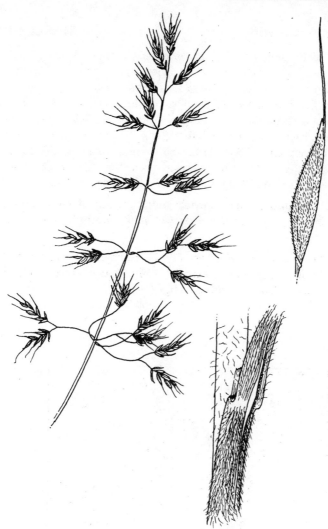

Bromus carinatus var. *carinatus*

Habitat and Distribution: Coastal prairies, grass balds, openings, chaparral, plains, and open oak and yellow pine woodlands. Variety *carinatus* occurs mainly west of the Cascade – Sierra Nevada summits from southeastern Vancouver Island to Baja California and in the Columbia River Gorge. Variety *hookerianus* occurs from California to Washington and eastward in the Columbia River Basin (to southeastern British Columbia and western Idaho), where intermediate forms interconnect it to var. *carinatus* in the west and to *B. marginatus* in the east.

Major References: Harlan (1945a, 1945b); Hitchcock & Chase (1951); Hooker & Arnott (1841); Shear (1900); Stebbins (1947, 1981); Stebbins & Tobgy (1944); Stebbins, Tobgy & Harlan (1944).

38. *Bromus catharticus* Vahl   RESCUE GRASS

Annual or biennial. Culms stout, solitary or tufted, erect or spreading, 30-120 cm tall; sheaths densely, often retrorsely short pubescent to villose or coarsely pilose, often with ciliate throat; ligules 1-4 mm long, erose, glabrous to pilose; without auricles; blades flat, 3-10 mm wide, glabrous to pilose; panicle 9-28 cm long, erect or nodding, open, with branches spreading or ascending; spikelets 20-30 mm long, with 6-12 florets, laterally compressed; first glume 7-12 mm long, 5-7 nerved; second glume 9-13 mm long, (7-)9(-11) nerved; lemmas compressed-keeled, 11-20 mm long, prominently (9-)11-13 nerved, glabrous to scabrous (or scabrous-pubescent) distally; awns 0-3.5 mm long; anthers to 2-4 mm long; $2n = 42$.

Habitat and Distribution: South America; widely introduced into North America as a forage crop, now adventive in disturbed soils; mostly in southern half of U.S.A.

Major References: Hitchcock & Chase (1951); Holmgren & Holmgren (1977); Hubbard (1956); Matthhei (1986); Parodi (1956); Pinto-Escobar (1976, 1981); Raven (1960); Shear (1900); Stebbins & Tobgy (1944).

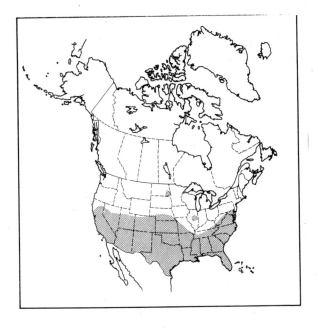

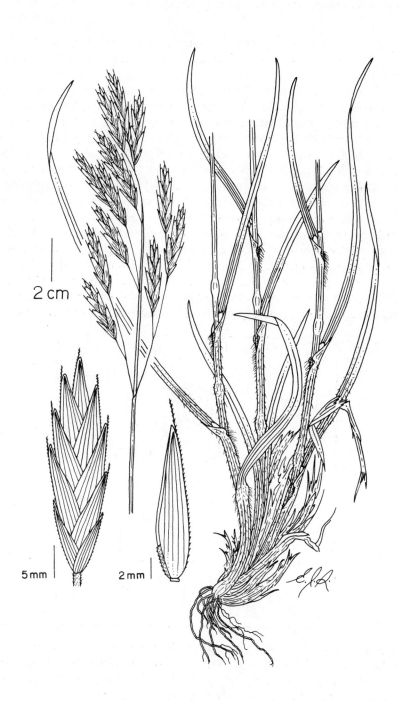

39. *Bromus subvelutinus* Shear. (*Bromus luzonensis* J. S. Presl in K. Presl)
HOARY BROME

Perennial. Tufted, with culms densely invested at base with old leaf sheaths, 45-80 cm tall; sheaths canescent (densely pilose, villose or short pubescent) usually with denser hairs near the collar; ligule 1-3 mm long, lacerate; lower leaves often with tiny auricles; blades narrow, 1-3 mm wide, 8-14 cm long, tending to become involute, canescent (pilose and/or short pubescent); panicle narrow, 5-15 cm long, the branches mostly simple, the lower sometimes compound, ascending-appressed; spikelets laterally compressed, 2-3 cm long; glumes pubescent (glabrous), the first (3-)5-7(-9) nerved, 8-11 mm long, the second 7-9(-11) nerved, 10-13 mm long; lemma 10-14(-17) mm long, antrorsely pubescent (scabrous), often distinctly 7-nerved, often with wide, hyaline margins; awn (3-)4-8(-10) mm long; anthers 4.5-6 mm long. 2n = 56. (*B. breviaristatus sensu* A.S. Hitchcock, *non* Buckl., in part. It is likely that *B. breviaristatus sensu* A. S. Hitchcock is the species referred to by Veldkamp (1990) as being the same species as *B. luzonensis*. Shear (1900) has pointed out that *B. breviaristatus* Buckl. is part of *B. marginatus*. I have examined the type specimens of both *B. subvelutinus* and *B. luzonensis* and conclude that they belong to the same species i.e. *B. subvelutinus sensu stricto*. Under the old ICBN rules the correct name for this species would be *B. luzonensis*, a name previously thought to be apply to a species from the Island of Luzon but now known to

apply to a western North American species whose type specimen was collected in California. Under the mandate of the XV International Botanical Congress not to disrupt established names for purely nomenclatural reasons, the established, taxonomically correct name of *B. subvelutinus* Shear is used here.)

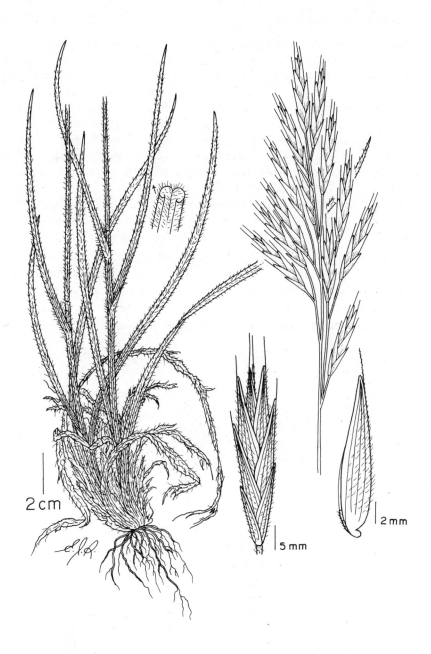

Habitat and Distribution: Grasslands, open pine or oak forests, 1,000-2,500 m elevation; California, western Nevada to northern Idaho.

Major References: Hitchcock & Chase (1951); Presl (1830); Shear (1900); Veldkamp (1990).

40. *Bromus marginatus* Nees   MOUNTAIN BROME

Perennial. Tufted; culms, 60-120(-180) cm tall, sometimes stout, puberulent or pubescent; sheaths typically sparsely retrorsely pilose throughout but ranging from densely pilose to pilose only at the throat; ligule 2-3.5 mm, obtuse, erose, sparsely hairy; blades flat, 15-25 cm long, 6-12 mm wide, sparsely pilose to pubescent on one or both sides or all glabrous; panicle mostly narrow, erect, 10-20(-30) cm long, with erect or ascending branches; spikelets 2.5-4mm long, laterally compressed, with 6-9 florets; glumes scabrous to pubescent, the first 7-9 mm long, 3-5 nerved, the second 9-11 mm long, 5-7 nerved; lemma coriaceous, 11-14 mm long, typically distinctly 7-9 nerved, pubescent on margins and back or on margins only or glabrous; awn 4-7 mm long; anthers 1-4 mm long. 2n = 56.

This is a variable species that is closely related to and has morphologically transitional forms with *B. carinatus* and *B. subvelutinus* on the west, *B. aleutensis* on the north and *B. polyanthus* on the southeast part of its range. Specimens with wider and taller i.e. more robust, culms, 20-30 cm long panicles with long lower branches (10-20 cm long) and 6-7 mm long awns have been called var. *latior* Shear. More leafy specimens that are generally less pubescent or more or less glabrous, and with glabrous glumes and scabrous to puberulent lemmas have been called var. *seminudus* Shear; these are morphologically close to *B. polyanthus*.

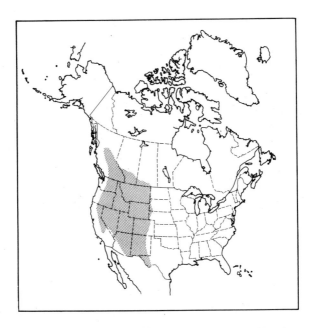

Habitat and Distribution:Open slopes, grass balds, shrublands, meadows and open forests (e.g. that of *Pinus ponderosa, Pinus contorta, Populus tremuloides*) in the montane and subalpine zones of the Cordillera and western Great Plains, at

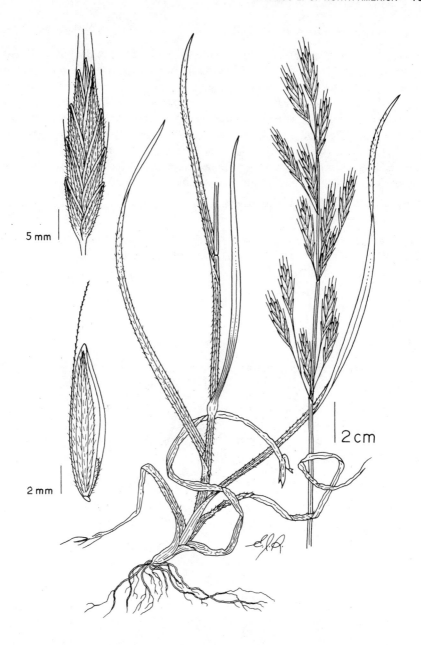

from 350-2,200 m elevation in the northern part of its range to 1,500-3,300 m elevation in the south. North-central British Columbia to south-central Saskatchewan, south to Arizona, west Texas, Mexico.

Major References: Hitchcock & Chase (1951); Shear (1900); Stebbins & Tobgy (1944).

41. *Bromus maritimus* **(Pip.) A.S. Hitchc.**   **MARITIME BROME**

Perennial. Culms stout, 22-70 cm tall, sometimes geniculate at the base and with numerous basal shoots; sheaths mostly glabrous or scaberulous, sometimes slightly pubescent at summit; ligules 1-6 mm long, acute or obtuse, erose, densely hairy to merely ciliolate; blades usually 6-8 mm wide, glabrous to scabrous; panicles 9-20 cm long, narrow, dense, with short, erect branches; spikelets laterally compressed, 3-4 cm long; glumes pubescent, the first (3-)5(-7) nerved, 8-12 mm long, the second 7(-9) nerved, 10-13 mm long; lemmas distinctly 9-11 nerved, 12-14 mm long, more or less uniformly antrorsely pubescent to pilose, often with bronze hyaline borders; awns (2-)4-7 mm long; anthers 2-4 mm long. 2n = 56.

Habitat and Distribution: Coastal sands. Pacific coast from Lane Co., Oregon to San Luis Obispo Co., California.

Major References: Hitchcock (1969); Hitchcock & Chase (1951); Munz (1968); Piper (1905); Stebbins (1981); Walters (1966); Wilkens & Painter (1993).

42. *Bromus polyanthus* Scribn. GREAT BASIN BROME
COLORADO BROME

Perennial. Culms stout, 60-120 cm tall, glabrous to puberulent; sheaths mostly glabrous and smooth, sometimes puberulent, pubescent or sparsely pilose; ligules 2-2.5 mm long, obtuse, erose, glabrous; blades scabrous to glabrous, rarely puberulent-pubescent near collar; panicle 15-25 cm long, the branches erect, ascending or spreading; spikelets 3-3.5 cm long, laterally compressed, 7-11 flowered; glumes smooth or scabrous, the first 3-nerved, the second 5-7 nerved; lemmas 12-15 mm long, 7-9 nerved, glabrous-scabrous, with broad, sometimes purplish hyaline margins; awn 4-8 mm long. 2n = 56. Closely related to *B. marginatus*. Sometimes separated into varieties as follows:

1. Panicle erect, somewhat contracted, elongate, with usually short, erect or ascending branches; awn 4-6 mm long .... var. *polyanthus*
1. Panicle nodding, open, wide with spreading branches; awn up to 8 mm long
    .... var. *paniculatus* Shear

Habitat and Distribution: Open slopes and meadows in the mountains. Montana to Washington, south to California and Texas.

Major References: Hitchcock & Chase (1951); Holmgren & Holmgren (1977); Munz (1968); Shear (1900); Stebbins (1981); Stebbins & Tobgy (1944).

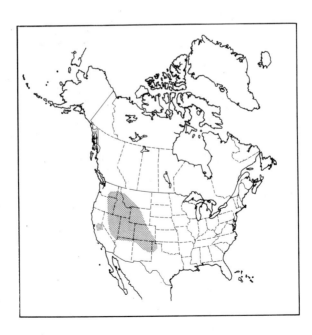

BROMUS L. OF NORTH AMERICA – 111

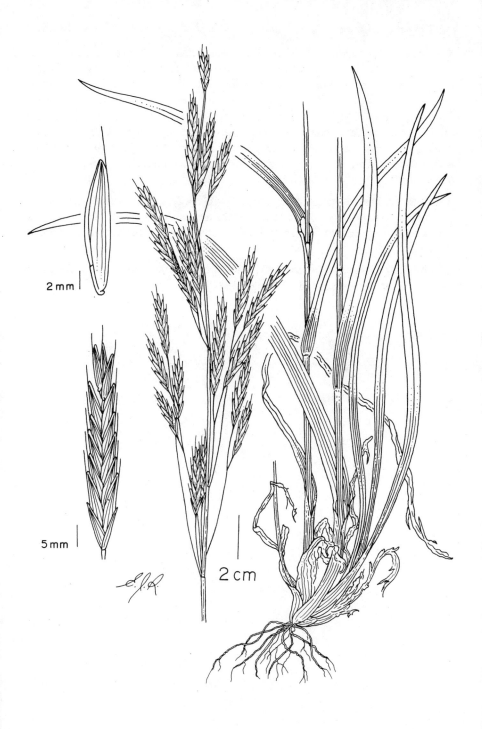

43. *Bromus sitchensis* Trin. in Bong. SITKA BROME, ALASKA BROME

Perennial. Culms stout, erect, 120-180 cm tall; sheaths glabrous to sparsely pilose; without auricles; ligule 3-4 mm long, obtuse, lacerate, glabrous or hairy; blades 20-40 cm long, sparsely pilose on upper or both sides; panicle open, 25-35 cm long, with the 2-4 lower branches up to 20 cm long, weak, spreading or patent, often drooping, with 1-3 spikelets born near the tips; glumes glabrous or scabrous, the first 3-5 nerved, the second 5-7 nerved; lemmas compressed-keeled, 12-14 mm long, 7-11 nerved, mostly glabrous or scabrous, sometimes hirtellous, sometimes very sparsely pilose along the margins; awn 5-10 mm long; anthers to 6 mm long. $2n = 42, 56$.

Habitat and Distribution: Exposed rock bluffs, cliffs, meadows and partial shade of forests along ocean edge, and road verges and other disturbed sites. Along the coast from southeastern Alaska to Washington. Closely related to *B. aleutensis* Trin. ex Griseb.

Major References: Bongard (1832); Hitchcock (1969); Hitchcock & Chase (1951); Shear (1900); Stebbins & Tobgy (1944).

44. *Bromus stamineus* Desv. in Gay   CHILEAN BROME

Annual. Culms often stout, 50-110 cm tall; sheaths densely retrorsely pilose, or sometimes densely villose generally or only on throat; ligules 1-3.5 mm long, thin, whitish, obtuse, lacerate to erose, glabrous; without auricles; leaf blades 3-10 mm wide, 7-30 cm long, glabrous to pilose or villose; panicle 10-25 cm long, mostly open and nodding, with branches stiffly spreading to ascending; spikelets laterally compressed, 2.8-4.2 cm long, with 4-6 florets; glumes glabrous to pubescent, the first 9-13 mm long, 5-7(-9) nerved, the second 11-17 mm long, 7-9-nerved; lemmas compressed-keeled, 10-16(-19) mm long, with 9-11 nerves that are prominent on distal half, usually pubescent at least near tip, with conspicuous whitish or partly purplish hyaline margins; awn (5-)8-12 mm long; anthers 0.6-1 mm long. 2n = 42. (Not included in A.S. Hitchcock.)

Habitat and Distribution: Disturbed soil, waste places, ballast. South America. Adventive in North America in central California and with records from northern Oregon and southern Washington.

Major References: Desvaux (1853); Moore (1983); Matthei (1986); Munz & Keck (1963); Nicora (1978); Wilken & Painter (1993).

Section *Genea*

45. *Bromus diandrus* Roth   **GREAT BROME**

Annual. Culms 35-90 cm, erect or decumbent; sheaths softly pilose, often with retrorse hairs ; ligule 2.5-3 mm long, obtuse, lacerate, glabrous; blades pilose both sides; panicle 13 -25 cm, lax, nodding, with lower branches spreading to patent, bearing 1 or 2 spikelet; spikelets 50-70 mm long, with callous scar nearly circular; glumes glabrous-scabrous with hyaline margins, the first 15-25 mm long, 3-nerved, the second 20-35 mm long, 5-nerved; lemma 20-35 mm long, 7-nerved, coarsely scabrous, lanceolate, hyaline bordered, long tapering to 3-4 mm long bifid apex; awn 35-65 mm, straight; anthers 0.5-1 mm. 2n = 56. (*B. rigidus* var. *gussonei* (Parl.) Coss. & Dur.).

Habitat and Distribution: Disturbed ground, waste places, fields. Southern Europe. Introduced to North America, now occurring in central and western California and recorded from Victoria, British Columbia and from Delaware. Closely related to *B. rigidus* and sometimes regarded as var. *gussonei* (Parl.) Coss. & Dur. of that species.

Major References: Gill & Carstairs (1988); Sales (1993); Smith (1970, 1980); Shear (1900).

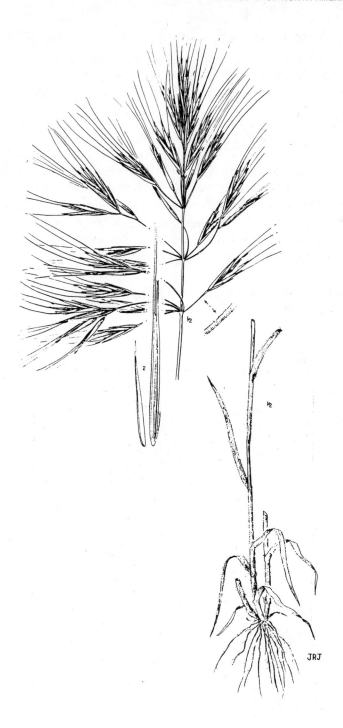

46. *Bromus madritensis* L.   MADRID BROME

Annual. Culms 34-70 cm tall, erect or ascending, mostly glabrous below panicle; lower sheaths densely short pubescent, the upper often smooth; ligules 1.5-2 mm long, obtuse, erose, glabrous; blades flat, pubescent both sides to glabrous; panicle loose, oblong-ovoid, 3-15 cm by 2-6 cm, erect, mostly with 2-3 ascending-spreading, visible branches at each node, with branches 10-30 mm long; spikelets reddish or purplish, not densely crowded, 30-50 mm long, wider at the top, lax with 6-10 florets; glumes pilose, hyaline bordered, the first 5-10 mm long, 1-nerved, the second 10-15 mm long, 3-nerved; lemma often 3 mm or more wide, 12-20 mm long, 5-7 nerved, linear-lanceolate, often arcuate, with hyaline margins ending in 1.5-3 mm apical teeth; awn 12-23 mm, straight or arcuate; anthers 0.5-1 mm. 2n = 14, 28, 42.

Habitat and Distribution: Disturbed soil, waste places, banks, road verges. Southern and western Europe. Introduced to North America and now occurring in west central California and recorded from the Reno, Nevada and Portland, Oregon areas.

Major References: Hitchcock & Chase (1951); Sales (1993); Shear (1900); Smith (1980); Wilken & Painter (1993).

47. *Bromus rigidus* Roth RIPGUT GRASS

Annual. Culms 20-53 cm tall, erect; sheaths softly pilose with retrorse or spreading hairs; ligules 2-3 mm long, obtuse, erose, glabrous; blades 10-25 cm long, pilose on both sides; panicle 13-20 cm long, more or less contracted, erect, with branches shorter than the spikelets; spikelets 25-35 mm long, erect with callous scar elliptical; glumes glabrous-scabrous with wide hyaline borders, the first 15-18 mm long, 3-nerved, the second 20-25 mm long, 5-nerved; lemma 22-28 mm long, 7-nerved, linear-lanceolate, with hyaline borders and the apex deeply bifid for 3-5 mm; awn 30-50 mm long, straight; anthers 0.8-1 mm long. 2n = 42.

Habitat and Distribution: Disturbed soil, waste places, fields, sand dunes, limestone areas. Southern and western Europe. Introduced to North America, and now occurring from southwestern British Columbia to Baja California and eastward to Idaho, Colorado, west Texas, and occasional in eastern USA.

Major References: Hitchcock (1969); Hitchcock & Chase (1951); Gill & Carstairs (1988); Sales (1993); Shear (1900).

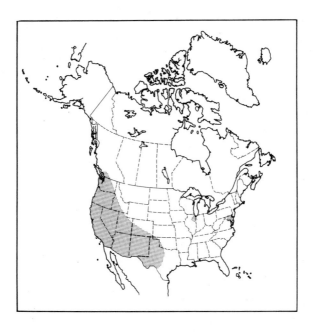

48. *Bromus rubens* **L.**   **FOXTAIL CHESS, RED BROME**

Annual. Culms 10-40 cm tall, erect or ascending, often puberulent below the panicle; sheaths softly pubescent to pilose; ligules 1-3 mm long, obtuse, lacerate, hairy; blades flat, to 12 cm long, pubescent both sides; panicle obovoid, 2-10 cm by 2-5 cm, densely contracted, cuneate at base, with branches 1-10 mm long, much shorter than the spikelets, and mostly not readily visible; spikelets densely crowded, often reddish-brown, 18-25 mm long, subsessile, with 4-8 florets; glumes with wide hyaline margins, pilose, the first 5-8 mm, 1(-3) nerved, the second 8-12 mm long, 3-5 nerved; lemma often 2-3 mm wide, 10-15 mm long, 7-nerved, pubescent-pilose, with hyaline margins ending in 1-3 mm apical teeth; awn 8-20 mm long, straight, reddish; anthers 0.5-1 mm long. $2n = 28$.

Habitat and Distribution: Disturbed ground, waste places, fields, rocky slopes. Southern and southwest Europe. Introduced to North America and occurring from southern Washington to southern California eastward to Idaho, Arizona, west Texas, and recorded from Massachusetts.

Major References: Hitchcock (1969); Hitchcock & Chase (1951); Smith (1980).

49. *Bromus sterilis* L. BARREN BROME

Annual. Culms 50-100 cm tall, glabrous, erect or geniculate near base; sheaths densely pubescent; ligules 2-2.5 mm long, acute, lacerate, glabrous; blades 5-20 cm long, pubescent on both sides; panicle 10-20 cm long, open, nodding, with patent, spreading or ascending branches that are mostly longer than the spikelets and which bear 1-3 spikelets; spikelets 20-35 mm long, with 5-9 florets; glumes glabrous-scabrous, the first 6-14 mm long, 1(-3) nerved, the second 10-20 mm long, 3(-5) nerved; lemmas pubescent-puberulent, 14-20 mm long, 7(-9) nerved, narrowly lanceolate, with 1-3 mm apical teeth; awn 15-30 mm, straight; anthers 1-1.4 mm long. $2n = 14, 28$.

Habitat and Distribution: Road verges, fields, waste places, overgrazed range. Europe from Sweden southward. Introduced to North America and now widespread in western and eastern North America, but mostly absent from the Great Plains region.

Major References: Hitchcock & Chase (1951); Munz & Keck (1963); Smith (1980); Wilken & Painter (1993).

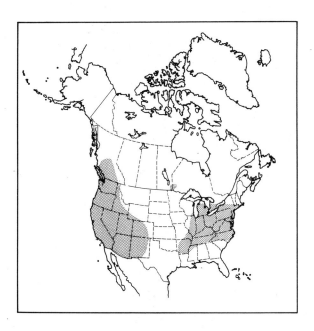

50. *Bromus tectorum* L.   **CHEATGRASS, DOWNY CHESS**

Annual. Culms 5-90 cm tall, erect, slender; sheaths densely, softly retrorsely pubescent to pilose or the upper ones sometimes glabrous; ligules 2-3 mm long, obtuse, lacerate, glabrous; blades 4-16 cm long, softly hairy on both sides; panicles 5-20 cm, more or less lax, erect at first then drooping and mostly secund, with slender flexuous branches, the longer of which bear 4-8 spikelets; spikelets 10-20 cm long, often purplish tinged; glumes villose-pubescent or glabrous, with hyaline margins, the first 4-9 mm long, 1-nerved, the second 7-13 mm long, 3-5 nerved; lemmas 9-12 mm long, 5-7 nerved, pubescent-pilose over the back and often with a few longer hairs on the margins, or all glabrous, lanceolate, with hyaline margins ending in 0.8-2(-3) mm long apical teeth; awn 10-18 mm long, straight; anthers 0.5-1 mm long. 2n = 14.

Habitat and Distribution: Disturbed sites (e.g., overgrazed rangeland, fields, sand dunes, road verges, waste places). Europe. Introduced into North America and now found throughout much of the USA and southern Canada. Specimens with glabrous spikelets may be referred to as forma *nudus* (Klett. & Richt.) St. John if such a distinction becomes necessary.

Major References: Hitchcock & Chase (1951); Morrow & Stahlman (1984); Shear (1900); Smith (1980); Upadhyaya, Turkington & McIlvride (1986).

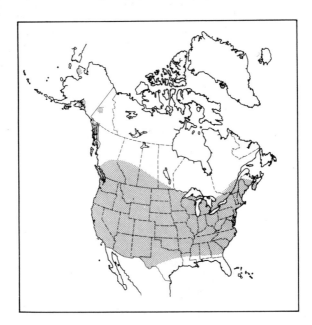

Section *Neobromus*

51. ***Bromus berterianus*** **Colla (*B. trinii* Desv. in Gay)**   **CHILEAN CHESS**

Annual. Culms slender, 30-60 cm tall; sheaths pilose-pubescent or nearly smooth; blades pilose to glabrous; panicle 10-20 cm long, erect, dense, narrow, with branches appressed to spreading, sometimes flexuous; spikelets 1.5-2(-2.5) cm long, narrow; glumes glabrous, acuminate, the first 8-11 mm long, 1(-3) nerved, the second 12-16 mm long, 3(-5) nerved; lemmas 11-14 mm long, acuminate, 5(-7) nerved, sparsely coarsely pubescent to pilose pubescent, with (1-)2-3 mm long aristate or acuminate teeth; awns (10-)13-20 mm long, proximally twisted and bent (var. *excelsus* not bent) and divaricate. 2n = 42. Specimens with 7-nerved lemmas, larger spikelets (2-2.5 mm long) and divaricate and somewhat twisted, but not bent awns have been called *B. trinii* var. *excelsus* Shear, and are known from the Panamint Mountains, California and Emory Canyon, Lake Mead, Arizona. With the acceptance of the name, *B. berterianus*, a new combination for this variety is needed and is here made. *Bromus berterianus* Colla var. *excelsus* (Shear) Pavlick, *comb. nov.* Basionym: *Bromus trinii* Desv. in Gay var. *excelsus* Shear. U.S. Dept. Agr. Div. Agrost. Bull. 23:25. 1900. Holotype: *Coville & Funston 522*, Panamint Mountains, Inyo Co., California, altitude 1,700 m.

Habitat and Distribution: Rocky slopes, dry canyons, plains and deserts. Chile, Argentina, western North America. Shear (1900), in his revision of North American species of *Bromus*, gave a general distribution for *B. trinii* as being from California to Colorado and south to Chile; Stebbins (1981) stated that this species

is native to the Pacific coast of both North and South America. It has been regarded by Hitchcock & Chase (1951) and in many recent works as being introduced to North America from Chile. In North America it occurs from southern Oregon to Baja California, eastward to Arizona and in Colorado and southwestern Utah.

Major References: Desvaux (1853); Hitchcock & Chase (1951); Matthei (1986); Munz & Keck (1963); Shear (1900); Stebbins (1981); Wilken & Painter (1993).

Excluded Species

1. *Bromus alopecuros* Poir.

Hitchcock & Chase (1951) recorded this species from Ann Arbor, Michigan; one specimen from there which may have been the basis for this record was previously identified to *B. alopecuros* but has since been re-identified to *B. hordeaceus* subsp. *divaricatus* (*B. molliformis*). Wilken & Painter (1993) reported it for California.

2. *Bromus laciniatus* Beal.

Hitchcock and Chase (1951) listed this species under section *Ceratochloa*, indicating that its range is Mexico and stated that it is occasionally cultivated for ornament. Soderstrom and Beaman (1968) listed *B. laciniatus* for Mexico as a synonym of *B. carinatus* Hook. & Arn., but did not indicate whether they had examined the type. Wagnon (1952, p. 470) stated that the type material of *B. laciniatus* is auriculate and belongs to *B. anomalus* of section *Bromopsis*. Stebbins (1981) agreed with Soderstrom & Beaman (1968) that this species should be united with *B. carinatus*.

3. *Bromus ramosus* Huds.

A specimen from Washington which may have been the basis on which *B. ramosus* was recorded for North America in Hitchcock & Chase (1951), was referred to *B. vulgaris* Hook. by C.L. Hitchcock (1969).

NOMENCLATURE

Section *BROMOPSIS*

1. *Bromus anomalus* Rupr. ex Fourn., Mex. Pl. 2:126. 1886. *B. anomalus* Rupr. in Galeotti, Bull. Acad. Roy. Brux. 9:236. 1842; *nom nudum*. *Zerna anomala* (Rupr. ex Fourn.) Henrard, Blumea 4:499. 1941. *Bromopsis anomala* (Fourn.) Holub, Folia Geobot. Phytotax., Praha 8:167. 1973. Lectotype selected by Soderstrom & Beaman (1968): Teotihuacan, Mexico, *Hahn* in 1865; P (lectotype); US, (fragment).
   B. ciliatus var. *minor* Munro ex Dewey, Contr. U.S. Natl. Herb. 2:548. 1894. Lectotype selected by Shear (1900): Chisos Mountains, West Texas, *Havard 20*; US.
   ? *B. laciniatus sensu* Wagnon, Brittonia 7:469, 470; *non* Beal, 1896.
   B. porteri var. *havardii* Shear, U.S. Dept. Agr., Div. Agrost. Bull. 23:37. 1900. Based on *B. ciliatus* var. *minor* Munro ex Dewey.
   B. meyeri Swallen, Contr. U.S. Nat. Herb. 29:395. 1950.

2. *Bromus ciliatus* L. Sp. Pl. 76.1753. *B. inermis* var. *ciliatus* (L.) Traut., Acta Hort. Petrop. 5:135. 1877. *B. hookeri* var. *ciliatus* (L.) Fourn., Mex. Pl. 2:128. 1886. *Forasaccus ciliatus* (L.) Lunell, Am. Midl. Nat. 4:225. 1915. *Zerna ciliata* (L.) Henrard, Blumea 4:498. 1941. *Bromopsis ciliata* (L.) Holub, Folia Geobot. Phytotax., Praha 8:167. 1973. Neotype: Essex County, New York, H. F. Heady 768. DAO; US.
   B. ciliaris Panzer, Pflanzensyst. 12:429, sphalm. 1775.
   B. canadensis Michx., Fl. Bor. Amer. 1:65. 1803. *B. hookeri* var. *canadensis* (Michx.) Fourn., Mex. Pl. 2:128. *Bromopsis canadensis* (Michx.) Holub, Folia Geobot. Phytotax., Praha 8:167. 1973. Canada, Lac St. Jean, *Michaux*; P (holotype); US, fragment.
   B. purgans var. *pallidus* Hook. Fl. Bor. Amer. 2:252. 1840. *B. richardsonii* var. *pallidus* (Hook.) Shear, U.S. Dept. Agr., Div. Agrost. Bull. 23:34. 1900. Saskatchewan to Rocky Mountains, *Drummond*.

B. ciliatus f. *denudatus* Wiegand, Rhodora 24:91. 1922. *B. ciliatus* var. *denudatus* (Wiegand) Fern., Rhodora 28:20. 1926. Ashfield, Mass., *Williams* in 1909. *B. ciliatus* subvar. *denudatus* (Wiegand) Farwell, Am. Midl. Nat. 10:204. 1927.

B. dudleyi Fern., Rhodora 32:63. 1930. Deer Brook, Bonne Bay, Newfoundland, *Fernald, Long & Fogg 1223*.

B. ciliatus var. *intonsus* Fern., Rhodora 32:70. 1930. Ashfield, Mass., *Williams* Aug. 4, 1909.

3. *Bromus erectus* Huds., Fl Angl. 39. 1762. *Festuca erecta* (Smith) -error for Huds.-Wallr., Sched. Crit. 35. 1822. *Bromopsis erectus* Fourr., Ann. Soc. Linn. Lyon 2, 17:187. 1869. *Zerna erecta* Panz. ex Jacks., Ind. Kew. 2:1249. 1895. *Forasaccus erectus* (Huds.) Bubani, Fl. Pyr. 4:384. 1901. England.

B. macounii Vasey, Torrey Bot. Club Bull. 15:48. 1888. Vancouver Island, *Macoun* in 1887.

4. *Bromus frondosus* (Shear) Woot. & Standl., New Mex. Exp. Sta. Bull. 81:144. 1912. *B. porteri* var. *frondosus* Shear, U.S. Dept. Agr., Div. Agrost. Bull. 23:37. 1900. *Bromopsis frondosa* (Shear) Holub, Folia Geobot. Phytotax., Praha 8:167. 1973. Type: Mangas, Grant County, New Mexico, *J. G. Smith* in 1897; US (holotype).

5. *Bromus grandis* (Shear) A. S. Hitchc. in Jeps., Fl. Calif. 3:175. 1912. *B. orcuttianus* var. *grandis* Shear, U.S. Dept. Agr., Div. Agrost. Bull. 23:43. 1900. *Bromopsis grandis* (Shear) Holub, Folia Geobot. Phytotax., Praha 8:167. 1973. Type: San Diego County, California, *C. R. Orcutt 472*: US.

B. porteri var. *assimilis* Davy, Univ. Calif. Publ. Bot. 1:55. 1902. San Jacinto Mountains, California, *Hall 2228*.

6. *Bromus inermis* Leyss., Fl. Halensis 16: 1761. *Festuca inermis* (Leyss.) DC. & Lam., Fl. Franc. 3:49. 1805. *Schedonorus inermis* Beauv., Ess. Agrost. 99:177. 1812. *Forasaccus inermis* (Leyss.) Lunell, Am. Midl. Nat. 4:225. 1915. *Zerna inermis* (Leyss.) Lindm., Svensk Fanerogamfl. 101. 1918. *Bromopsis inermis* (Leyss.) Holub, Folia Geobot. Phytotax., Praha 8:167. 1973. Europe.

Festuca inermis var. *villosa* Mert. & Koch, Deutschl. Fl. 1:675. 1823. *B. inermis* f. *villosus* (Mert. & Koch) Fern., Rhodora 35:316. 1933. Germany.

B. inermis var. *aristatus* Schur., Enum. Pl. Transsilv. 805. 1866. *B. inermis* f. *aristatus* (Schur.) Fern., Rhodora 35:316. 1933. Europe.

B. inopinatus Brues, Trans. Wis. Acad. Sci. 17:73. 1911. Milwaukee, Wisconsin, *Brues 78*.

B. inermis f. *bulbiferus* Moore, Rhodora 43:76. 1941. Ramsey County, Minnesota, *Kaufman* in 1938.

7. *Bromus kalmii* A. Gray, Man. Bot. 600. 1848. *Bromopsis kalmii* (A. Gray) Holub, Folia Geobot. Phytotax., Praha 8:167. 1973.
   B. purgans L., Sp. Pl. 76. 1753, nom. rejic. (Art. 69). *B. imperialis* Steud., Nom. Bot. ed. 2, 1:229. 1840; synonym. *B.steudelii* Frank ex Steud., Nom. Bot. ed. 2, 1:229. 1840; synomym. *Forasaccus purgans* (L.) Lunell, Am. Midl. Nat. 4:225. 1915. *Zerna purgans* (L.) Henrard, Blumea 4:498. 1941. Lectotype selected by A. S. Hitchcock: Saratoga to Fort St. Frederic, New York, 24 June - 1 July, 1749, *P. Kalm*; LINN.

8. *Bromus laevipes* Shear, U.S. Dept. Agr., Div. Agrost. Bull. 23:45. 1900.
   Bromopsis laevipes (Shear) Holub, Folia Geobot. Phytotax., Praha 8:168. 1973. Columbia River, West Klickitat County, Washington, *W. N. Suksdorf 178*; US.

9. *Bromus lanatipes* (Shear) Rydb., Colo. Agr. Exp. Sta. Bull. 100:52. 1906. *B. porteri* var. *lanatipes* Shear, U.S. Dept. Agr., Div. Agrost. Bull. 23:37. 1900. *B. anomalus* var. *lanatipes* (Shear) A. S. Hitchc., Wash. Acad. Sci. Jour. 23:449. 1933. *Bromopsis lanatipes* (Shear) Holub, Folia Geobot. Phytotax., Praha 8:168. 1973. Idaho Springs, Colorado, *C.L.Shear 739*; US (holotype).
   B. lanatipes f. *glaber* Wagnon, Leafl. West. Bot. 6:68. 1950. Holotype: Santa Fe County, New Mexico, *A. A. & E. Gertrude Heller 3835*; NY.

10. *Bromus latiglumis* (Scribn. ex Shear) A. S. Hitchc., Rhodora 8:211. 1906. *B. purgans* var. *latiglumis* Scribn. ex Shear, U.S. Dept. Agr., Div. Agrost. Bull. 23:40. *Forasaccus latiglumis* (Scribn. ex Shear) Lunell, Am. Midl. Nat. 4:225. 1915. *B. ciliatus* var. *incanus* subvar. *latiglumis* (Scribn. ex Shear) Farwell, Am. Midl. Nat. 10:204. 1927. *Zerna latiglumis* (Scribn. ex Shear) Henrard, Blumea 4:498. 1941. Dakota City, Iowa, *Pammel 222*.
    B. altissimus Pursh, Fl. Amer. Sept. 2:728. 1814, *non* Weber ex Wiggers, 1780; *non* Gilib., 1792.
    B. ciliatus var. *latiglumis* Scribn. ex Shear, U.S. Dept. Agr., Div. Agrost. Bull. 23:40. 1900; as synonym of *B. purgans* var. *latiglumis* Scribn. ex Shear.
    B. purgans var. *incanus* Shear, U.S. Dept. Agr., Div. Agrost. Bull. 23:41. 1900. *B. incanus* (Shear) A. S. Hitchc., Rhodora 8:212. 1906. *B. altissimius* f. *incanus* (Shear) Wiegand, Rhodora 24:91. 1922. *B. ciliatus* var. *incanus* (Shear) Farwell, Am. Midl. Nat. 10:204. 1927. *B. latiglumis* f. *incanus* (Shear) Fern., Rhodora 35:316. 1933. Canton, Illinois, *J. Wolf 3*.
    B. purgans sensu Wagnon, Brittonia 7:452. 1952, *non* L., 1753.
    B. ciliatus sensu Baum, Can. J. Bot. 45:1849. 1967, *non* L., 1753.

11. *Bromus mucroglumis* Wagnon, Leafl. West. Bot. 6:67-68. 1950. *Bromopsis macroglumis* (Wagnon) Holub, Folia Geobot. Phytotax., Praha 8:168, sphalm. 1973. Holotype: *H. K. Wagnon 1520;* grown from seed produced at Albuquerque, New Mexico, the original source of which was Chiricahua Mountains, Cochise County, Arizona, collected by Goodding, Locke & Johnson; MICH.

12. *Bromus nottowayanus* Fern., Rhodora 43:530. 1941. *Bromopsis nottowayana* (Fern.) Holub, Folia geobot. phytotax., Praha 8:168. 1973. Sussex County, Virginia, *M. L. Fernald & B. Long 12239*; GH.

13. *Bromus orcuttianus* Vasey, Bot. Gaz. 10:223. 1885. *Bromopsis orcuttiana* (Vasey) Holub, Folia Geobot. Phytotax., Praha 8:168. 1973. Type: San Diego, California, *C. R. Orcutt* in 1884; US.
    B. brachyphyllus Merr., Rhodora 4:146. 1902. Crook County, Oregon, *Cusick 2677*.
    Bromus orcuttianus var. *hallii* A. S. Hitchc. in Jeps., Fl. Calif. 1:175. 1912. Type: San Jacinto Mountains, Riverside County, California, *H. M. Hall 2301*; US.

14. *Bromus pacificus* Shear, U.S. Dept. Agr., Div. Agrost. Bull. 23:38. 1900. *Bromopsis pacifica* (Shear) Holub, Folia Geobot. Phytotax., Praha 8:168. 1973. Type: Seaside, Clatsop County, Oregon, *Scribner & Shear 1703*; US.
    B. magnificus Elmer, Bot. Gaz. 36:53. 1903. Port Angeles, Washington, *Elmer 1957*.

15. *Bromus porteri* (Coult.) Nash, Torrey Bot. Club Bull. 22:512. 1895. *B. kalmii* var. *porteri* Coult., Man. Bot. Rocky Mts. Reg. 425. 1885. *B. ciliatus* var. *porteri* (Coult.) Rydb., Contr. U.S. Natl. Herb. 3:192. 1895. *Bromopsis porteri* (Coult.) Holub, Folia Geobot. Phytotax., Praha 8:168. Type: Twin Lakes, Lake County, Colorado, T. C. Porter in 1872; NY (lectotype selected by Wagnon, 1952).
    B. kalmii var. *occidentalis* Vasey ex Beal, Grasses N. Amer. 2:624. 1896. Type: Montanus, *Canby & Scribner 384*.
    B. ciliatus var. *montanus* Vasey ex Beal, Grasses N. Amer. 2:619. 1896; sphalm., var. *montanis*. *B. ciliatus* var. *montanus* Vasey, Bot. Wheeler Exped. 292. 1878; *nom. nudum*. Type: Colorado, *Patterson 264;* MSC (holotype), US.
    B. ciliatus var. *scariosus* Scribn., U.S. Dept. Agr. Div. Agrost. Bull. 13:46. 1898.
    B. scabratus Scribn. U.S. Dept. Agr., Div. Agrost. Bull. 13:46. 1898, *non* Link, 1843. Vermilion Creek, Wyo., *A. Nelson 3800*.
    B. kalmii var. *major* Vasey ex Shear, U.S. Dept. Agr., Div. Agrost. Bull. 23:35. 1900; as synonym of *B. porteri* (Coult.) Nash.

16. *Bromus pseudolaevipes* Wagnon, Leafl. West. Bot. 6:64-65. 1950. *Bromopsis pseudolaevipes* (Wagnon) Holub, Folia Geobot. Phytotax., Praha 8:168. 1973. Holotype: *H. K. Wagnon 1507*, grown from seed originally collected by G. L. Stebbins, Jr. in Los Angeles County, California; MICH.

17. *Bromus pubescens* Muhl. ex Willd., Enum. Pl. Hort. Berol. 120. 1809. *B. hookeri* var. *pubescens* (Muhl. ex Willd.) Fourn., Mex. Pl. 2:127. 1886. *Bromopsis pubescens* (Willd.) Holub, Folia Geobot. Phytotax., Praha 8:168. 1973. Type: Lancaster County, Pennsylvania, *Muhlenberg M154*; PH.
    B. purgans auct. plur., *non* L., 1753. *B. ciliatus* var. *purgans* (L.) A. Gray, Man. Bot. 600. 1848.
    B. ciliatus var. *laeviglumis* Scribn. ex Shear, U.S. Dept. Agr., Div. Agrost. Bull. 23:32. 1900. *Forasaccus ciliatus* var. *laeviglumis* (Scribn. ex Shear) Lunell, Am. Midl. Nat. 4:225. 1915. *B. ciliatus* f. *laeviglumis* (Scribn. ex Shear) Wiegand, Rhodora 24:91. 1922. *B. laeviglumis* (Scribn. ex Shear) A. S. Hitchc., Proc. Biol. Soc. Wash. 41:157. 1928. *B. purgans* var. *laeviglumis* (Scribn. ex Shear) Swallen, Proc. Biol. Soc. Wash. 54:45. 1941. Galt, Ontario, *Herriot* in 1898.
    B. purgans f. *glabriflorus* Wiegand, Rhodora 24:92. 1922. Ithaca, New York, *Metcalf 5813*.
    B. purgans f. *laevivaginatus* Wiegand, Rhodora 24:92. 1922. *B. ciliatus* var. *purgans* subvar. *laevivaginatus* (Wiegand) Farwell, Am. Midl. Nat. 10:204. 1927. Ithaca, New York, *Metcalfe 5821*.

18. *Bromus pumpellianus* Scribn., Torrey Bot. Club Bull. 15:9. 1888. *Forasaccus pumpellianus* (Scribn.) Lunell, Am. Midl. Nat. 4:225. 1915. *B. inermis* subsp. *pumpellianus* (Scribn.) Wagnon, Rhodora 52:211. 1950. *Zerna pumpelliana* (Scribn.) Tzvelev, Fl. Arct. U. R. S. S. 2:225. 1964. *Bromopsis pumpelliana* (Scribn.) Holub, Folia Geobot. Phytotax. 8:168. 1973. Belt Mountains, Meagher County, Montana, *F. Lamson-Scribner 418*; US.
    B. purgans var. *purpurascens* Hook., Fl. Bor. Amer. 2:252. 1840. *B. inermis* subsp. *pumpellianus* var. *purpurascens* (Hook.) Wagnon, Rhodora 52:211. 1950. Bear Lake to Arctic seacoast, *Richardson*.
    B. purgans var. *longispicatus* Hook., Fl. Bor. Amer. 2:252.1840. Rocky Mountains, *Drummond*.
    ? *B. purgans* Hook. & Arn., Bot. Beechey Voy. Suppl. 132. 1841.
    B. ciliatus var. *coloradensis* Vasey ex Beal, Grasses N. Amer. 2:619. 1896. Lectotype selected by A. S. Hitchcock: near Gray's Peak, Colorado, *Wolf 1158*.
    B. pumpellianus var. *tweedyi* Scribn. ex Beal, Grasses N. Amer. 2:622. 1896. Yellowstone Park, *Tweedy 587*.

B. pumpellianus var. *melicoides* Shear, U.S. Dept. Agr., Div. Agrost. Bull. 23:50. 1900. Beaver Creek Camp, Colorado, *Pammel* in 1896.

B. arcticus Shear in Scribn. & Merr., Contr. U.S. Natl. Herb. 13:83. 1910. *B. pumpellianus* var. *arcticus* (Shear) Porsild, Rhodora 41:182. 1939.

B. pumpellianus var. *villosissimus* Hulten, Lunds Univ. Arsskr. 2, Sect. 2, 38:251. 1942.

B. pumpellianus subsp. *dicksonii* Mitch. & Wilt., Brittonia 18:163. 1966. Holotype: Yukon River, Alaska, *J. G. Dickson LY 57-19*. ALA.

19. *Bromus richardsonii* Link, Hort. Berol. 2:28. 1833. *Zerna richardsonii* (Link) Nevski, Act. Univ. Asiae Med. 8b. Bot. 17:17. 1934. *Bromopsis richardsonii* (Link) Holub, Folia Geobot. Phytotax., Praha 8:168. 1973. Grown at Berlin from seed collected in northwestern North America; *Richardson*.

20. *Bromus suksdorfii* Vasey, Bot. Gaz. 10:223. 1885. *Bromopsis suksdorfii* (Vasey) Holub, Folia Geobot. Phytotax. 8:169 Type: Mount Adams, Washington, *W. N. Suksdorf 74*; US.

21. *Bromus texensis* (Shear) A. S. Hitchc., Contr. U.S. Natl. Herb. 17:381. 1913. *B. purgans* var. *texensis* Shear, U.S. Dept. Agr., Div. Agrost. Bull. 23:41. 1900. *Bromopsis texensis* (Shear) Holub, Folia Geobot. Phytotax. 8:169. Bexar County, Texas, *Jermy 230*; US.

22. *Bromus vulgaris* (Hook.) Shear, U.S. Dept. Agr., Div. Agrost. Bull. 23:43. 1900. *B. purgans* var. *vulgaris* Hook., Fl. Bor. Amer. 2:252. 1840; as to type only. *Zerna vulgaris* (Hook.) Henrard, Blumea 4:498. 1941. *Bromopsis vulgaris* (Hook.) Holub, Folia Geobot. Phytotax. 8:169. 1973. Lectotype selected by Wagnon (1952): Columbia River, *Dr. Scouler*; K.

B. ciliatus var. *ligulatus* Vasey ex Macoun, Cat. Can. Pl. 2:238. 1888; nomen nudum; Vancouver Island, *Macoun* in 1887.

B. ciliatus var. *pauciflorus* Vasey ex Macoun, Cat. Can. Pl. 2:238. 1888; nomen nudum; Oregon, *Howell*.

B. debilis Nutt. ex Shear, U.S. Dept. Agr., Div. Agrost. Bull. 23:43. 1900; as synonym of *B. vulgaris*.

B. vulgaris var. *eximius* Shear, U.S. Dept. Agr., Div. Agrost. Bull. 23:44. 1900. *B. eximius* (Shear) Piper, Contr. U.S. Natl. Herb. 11:143. 1906. Wallowa Lake, Oregon, *Shear 1791*.

B. vulgaris var. *robustus* Shear, U.S. Dept. Agr., Div. Agrost. Bull. 23:44. 1900. Seaside, Oregon, *Scribner & Shear 1710*.

B. ciliatus var. *glaberrimus* Suksdorf, Deut. Bot. Monatsschr. 19:93. 1901. Skamania County, Washington, *Suksdorf* in 1894 (2335).

B. eximius var. *umbraticus* Piper, Contr. U.S. Natl. Herb. 11:144. 1906. Based on *B. vulgaris* Shear.

Section *BROMUS*

23. *Bromus arenarius* Labill., Nov. Holl. Pl. 1:23. 1804. Australia.

24. *Bromus arvensis* L., Sp. Pl. 77. 1753. *B. erectus* var. *arvensis* Huds., Fl. Angl. ed. 2, 50. 1778. *Serrafalcus arvensis* Godr., Fl. Lorr. 3:185. 1844. *Forasaccus arvensis* Bubani, Fl Pyr. 4:385. 1901. Europe.

25. *Bromus briziformis* Fisch. & C. A. Mey. in Fisch., Ind. Sem. Hort. Petrop. 3:30, as *B. brizaeformis,* orthographic error (I. C. B. N. Art. 73.8). 1837. Europe.

26. *Bromus commutatus* Schrad., Fl. Germ. 353. 1806. *Brachypodium commutatum* Lam., sphalm. for Schrad.) Beauv., Ess. Agrost. 101, 155. 1812. *Serrafalcus commutatus* Bab., Man. Brit. Bot. ed. 1, 374. 1843. *B. mutabilis* var. *commutatus* (Schrad.) Schultz, Flora 32:234. 1849. *B. racemosus* var. *commutatus* Coss. & Dur., Eppl. Sci. Alger. 2:165. 1855. *B. mollis* var. *commutatus* (Schrad.) Sanio, Verh. Bot. Ver. Brand. 23: Abh. 31. 1882. *Serrafalcus racemosus* var. *commutatus* (Schrad.) Husnot, Gram. Fr. Belg. 72. 1899. *Forasaccus commutatus* Bubani, Fl. Pyr. 4:387. 1901. *B. racemosus* subsp. *commutatus* (Schrad.) Tourlet, Catal. Pl. Vasc. Indre-et-Loire, 588. 1908. Germany.
B. pratensis Ehrh., Beitrage 6:84. 1791; *nomen nudum*; Hoffm., Deut. Fl. ed. 2, 2:52. 1800., *non* Lam., 1785. *B. mutabilis* var. *pratensis* (Ehrh.) Schultz, Flora 32:234. 1849. *B. hordeaceus* var. *pratensis* (Ehrh.) Fiori, Nuova Fl. Analitica Ital. 1:149. 1923. Europe.
B. secalinus var. *gladewitzii* Farwell, Am. Midl. Nat.. 10:24. 1926. Michigan, *Farwell & Gladewitz 7434.*
B. commutatus var. *apricorum* Simonkai, Enum., Fl. Transsilv. 583. 1886. *B. pratensis* var. *apricorum* (Simonkai) Druce, Fl. Oxfordshire, ed. 2. 496. 1927. Europe.
B. commutatus subsp. *neglectus* (Parl.) P. M. Smith, Bot. J. Linn. Soc. 76:360. 1978. *Serrafalcus neglectus* Parl., ?

27. *Bromus hordeaceus* L., Sp. Pl., ed. 1:77. 1753.
subsp. *hordeaceus.*
B. mollis L., Sp. Pl. ed. 2. 1:112. 1762. *Serrafalcus mollis* (L.) Parl., Pl. Rar. Sic. 2:11. 1840. *Forasaccus mollis* (L.) Bubani, Fl Pyr. 4:386.1901. *B.*

hordeaceus var. *mollis* (L.) Fiori, Nuova Fl. Analitica Ital. 1:149. 1923. *B. hordeaceus* subsp. *mollis* Maire in Emberger & Maire, Cat. Pl. Maroc 4:943. 1941. Europe.

subsp. *divaricatus* (Bonnier & Layens) Kerguelen, Bull. Soc. Ech. Pl. Vasc. Eur. Bass. Medit. (Liege), 18:27. 1981. *B. intermedius* Guss. subsp. *divaricatus* Bonnier & Layens, Tabl. Syn. Pl. Vasc. Fl. Fr., 369. 1894.

B. divaricatus sensu Lloyd, Fl. Loire-Inf., 314. 1844; *non* Rohde in Loisel., Notice, 22. 1810.

B. hordeaceus subsp. *molliformis* (Lloyd ex Godron) Maire & Weill. in Maire, Fl. Afr. Nord 3:255. 1955. *B. molliformis* Lloyd ex Godron, Fl. Loire-Inf. 315. 1844. *Serrafalcus molliformis* (Lloyd ex Godron) F. W. Schultz, Arch. Fl. Fr. Allem. :320. 1861. Type: Pornic-Saint-Brevin (France), *Lloyd*.

B. confertus Boreau, Fl. Centre Fr., 2:586. 1849.; *non* Bieb., 1808.

Serrafalcus lloydianus Gren. & Godron, Fl. Fr., 3:591. 1855. *B. lloydianus* (Gren. & Godron) Nyman, Syll. Suppl. :73. 1865.

subsp. *pseudothominei* (*"pseudothoninii"*) (P. Smith) H. Scholz, Willdenowii 6:148. 1970. *B. x pseudothominei* P. Smith, Watsonia 6:330. 1968.

B. mollis var. *leiostachys* Hartm., Skand. Fl. Handb. ed. 2, 33. 1832; *pro parte*. *B. mollis* f. *leiostachys* (Hartm.) Fern., Rhodora 35:316. 1933; *pro parte*. Sweden.

B. gracilis var. *micromollis* Krosche, Feddes Rep. 19, 329. 1924; *pro parte*.

B. lepidus f. *lasiolepis* Holmb., Bot. Not. 1924, 326. 1924; *pro parte*.

B. thominei sensu Tutin in Clapham *et al*, Fl. Brit. Isles ed. 2, 1152. 1962; *non* Hardouin, 1833.

subsp. *thominei* (Hard.) Maire in Emberger & Maire, Catal. Pl. Maroc, 4:943. 1941; nomenclaturally superfluous when published but legitimate; correct when subsp. *thominei* is recognized as distinct from subsp. *hordeaceus* (Art. 63.3); authorship has been cited as (Hardouin (=Ardoino)) Hylander, Upps. Univ. Arsskr., 1945:84. 1945, or (Hardouin) Maire & Weill. in Maire, Fl. Afr. Nord 3:256. 1955. *B. thominei (*"thominii"*)* Hard., Congres. Sc. Fr. 1, 56. 1833; *non sensu* Tutin in Clapham *et al*., Fl. Brit. Isles, ed. 2, 1152. 1962.

? *B. nanus* Weigel, Obs. Botan. 8. 1772.

B. arenarius Thomine-Desmazures, Mem. Soc. Linn. Calvados 1824, 40. 1824; *non* Labill., 1804.

28. **Bromus japonicus** Thunb. in Murray, L. Syst. Veg., ed. 14, 119. 1784; Thunb., Fl. Jap. 52. 1784; serius. Type: Japan.

B. vestitus Schrad., Goetl. Anz. Ges. Wiss. 2:2074. 1822.

B. patulus Mert. & Koch, Deut. Fl. 1:685. 1823. *B. arvensis* var. *patulus* (Mert. & Koch) Mutel, Fl. Franc. 4:134. 1837. *Serrafalcus patulus* (Mert. &

Koch) Parl., Fl. Ital. 1:394. 1848. *B. squarrosus* var. *patulus* (Mert. &
Koch) Regel, Act. Hort. Petrop.7:602.1881. *Forasaccus patulus* (Mert. &
Koch) Bubani, Fl. Pyr. 4:387. 1901. Europe.

B. cyri Trin. in C. A. Meyer, Verz. Pfl. Cauc.:24. 1831.

B. pendulus and *B. unilateralis* Schur., Enum. Pl. Transs.:802. 1866. *B. unilateralis pro syn. B. pendulus.*

B. chiapporianus de Not. ex Nyman, Consp.:824. 1882.; *pro syn B. patulus* Mert. & Koch.

B. japonicus var. *porrectus* Hack., Magyar Bot. Lapok (Ungar. Bot. Bl.) 2:58. 1903. Eurasia.

B. japonicus var. *subsquarrosus* (Borb.) Savul. & Rays., (Rumania) Min. Agr. Bull. 4, Sup. 2:39. 1924; as synonym of *B. japonicus* var. *porrectus* Hack.

B. abolinii Drob., Feddes Repert., 21:40. 1925.

B. gedrosianus Penzes, Bot. Kozl., 33:111. 1936.

B. japonicus subsp. *anatolicus* (Boiss. & Heldr.) Penzes, Bot. Kozl. 33:118. 1936.

B. multiflorus auct. *non* Weigel: DC. in Lam. & DC., Fl. Fr., ed. 3, 3:69. 1805.

29. *Bromus lanceolatus* Roth, Catalecta Bot. 1:18. 1797.
B. macrostachys Desf., Fl. Atlant. 1:96, pl.19, f. 2. 1798. *Serrafalcus macrostachys* (Desf.) Parl., Fl. Ital. 1:397. 1848. *Zerna macrostachys* Panz. ex Jacks., Ind. Kew. 2:1249. 1895. Algeria.

30. *Bromus lepidus* Holmb., Bot. Not. 1924, 326. 1924.
B. gracilis Krosche, Feddes Rep. 19, 329. 1924; *non* Leyss., 1761.
B. brittanicus I. A. Williams, J. Bot. Lond., 67, 65. 1929.

31. *Bromus racemosus* L., Sp. Pl. ed. 2, 1:114. 1762. *Serrafalcus racemosus* (L.) Parl., Rar. Pl. Sic. 2:14. 1840. *B. arvensis* var. *racemosus* (L.) Neilreich, Fl. Nieder-Oesterr. 81. 1859. *B. squarrosus* var. *racemosus* (L.) Regel, Act. Hort. Petrop. 7:602. 1881. *B. mollis* var. *racemosus* (L.) Fiori, Fl. Analitica Ital. 1:100. 1896. *Forasaccus racemosus* (L.) Bubani, Fl. Pyr. 4:387. 1901. *B. hordeaceus* var. *racemosus* (L.) Fiori, Nuova Fl. Analitica Ital. 1:149. 1923. Europe.

32. *Bromus scoparius* L., Cent. Pl. 1:6. 1755; Amoen. Acad. 4:266. 1759.
Serrafalcus scoparius (L.) Parl., Fl. Palerm. 1:174. 1845. Spain.

33. *Bromus secalinus* L., Sp. Pl. 76. 1753. *B. mollis* var. *secalinus* (L.) Huds., Fl. Angl. ed. 2, 49. 1778. *Avena secalinus* (L.) Salisb., Prodr. Stirp. 22. 1796. *Serrafalcus secalinus* (L.) Bab., Man. Brit. Bot. ed. 1. 374. 1843. *Forasaccus secalinus* (L.) Bubani, Fl. Pyr. 4:388. 1901. Europe.
?*B. submuticus* Steud., Syn. Pl. Glum. 1:321. 1854. St. Louis, Mo.

B. secalinus var. *velutinus* Koch, Syn. Fl. Germ. Helv. 819. 1837. *B. velutinus* Schrad., Fl. Germ. 1:349. pl.6. f. 3. 1806. Germany.

34. *Bromus squarrosus* L., Sp. Pl. 1:76. 1753. France, Switzerland, Siberia.
B. wolgensis Fisch. ex Jacq...

Section *CERATOCHLOA*

35. *Bromus aleutensis* Trin. ex Griseb. in Ledeb., Fl. Ross. 4:361. 1853. *B. sitchensis* var. *aleutensis* (Griseb.) Hulten, Fl. Alas. 254. 1942. Unalaska, *Eschscholtz*.

36. *Bromus arizonicus* (Shear) Stebbins, Calif. Acad. Sci. 4. Proc. 25:309. 1944. *B. carinatus* var. *arizonicus* Shear, U.S. Dept. Agr., Div. Agrost. Bull. 23:62. 1900. *Ceratochloa arizonica* (Shear) Holub, Folia Geobot. Phytotax. 8:170. 1973. Santa Cruz Valley, Tucson, Arizona, *Pringle* in 1884.

37. *Bromus carinatus* Hook. & Arn., Bot. Beechey Voy. Suppl. 403. 1841.
Ceratochloa carinata (Hook. & Arn.) Tutin in Clap., Tutin and Warb., Fl. Brit. Is. 1458. 1952. Type: Monterey or San Francisco, California, "at no great distance from the coast", *D. Douglas* in 1833; BM (lectotype!); US (fragment).
Ceratochloa grandiflora Hook., Fl. Bor. Amer. 2:253. 1840, *non B. grandiflorus* Weigel, 1772. *Bromus hookerianus* Thurb. in Wilkes, U.S. Expl.Exped. Bot. 17:493. 1874. *B. carinatus* var. *hookerianus* (Thurb.) Shear, U.S. Dept. Arg., Div. Agrost. Bull. 23:60. 1900. Plains of the Columbia (Oregon), *Scouler, Douglas*.
Bromus oregonus Nutt. ex Hook. f., J. Bot. Kew. Misc. 8:18. 1856, nomen; Nutt. ex Shear, U.S. Dept. Agr., Div. Agrost. Bull. 23:59. 1900, in synonymy. "Upper Missouri and Oregon territories", *Geyer* 244.
Bromus virens Buckl., Acad. Nat. Sci. Phila. Proc. 1862:98. 1862. "Rocky Mountains and Columbia River, *Nuttall*."
Bromus californicus Nutt. ex Buckl., Acad. Nat. Sci. Phila. Proc. 1862:336. 1862, as synonym of *B. virens* Buckl. *B. carinatus* var. *californicus* (Nutt.) Shear, U.S. Dept. Agr., Div. Agrost. Bull. 23:60..1900. California, *Nuttall*.
Bromus nitens Nutt. ex A. Gray, Acad. Nat. Sci. Phila. Proc. 1862:336. 1862, as synonym of *B. virens* Buckl. Columbia woods, *Nuttall*.
Bromus carinatus var. *linearis* Shear, U.S. Dept. Agr., Div. Agrost. Bull. 23:61.1900. California, *Vasey* in 1875.
Bromus hookerianus var. *minor* Scribn. ex Vasey, Descr. Cat. Grasses U.S. 92. 1885, nomen. Oregon.

Bromus virens var. *minor* Scribn. in Beal, Grasses N. Amer. 2:614. 1896.
Arizona and Oregon.

Bromus carinatus var. *densus* Shear, U.S. Dept. Agr., Div. Agrost. Bull. 23:61. 1900. San Nicolas Island, California, *Trask* 12.

38. *Bromus catharticus* Vahl, Symb. Bot. 2:22. 1791. *Ceratochloa cathartica* Herter, Rev. Sudamer. Bot. 6:144. 1940. Types: Lima, Peru, *Joseph Dombey*; P-JU (lectotype selected by Pinto-Escobar (1976); P (isotype).

B. unioloides Kunth in HBK, Nov. Gen. Sp. 1:151. 1815. *Schedonorus unioloides* (Kunth) Roem. & Schult., Syst. Veg. 2:709. 1817. *B. mucronatus* Willd. ex Steud., Nom. Bot. ed. 2. 1:228. *Zerna unioloides* (Kunth) Lindm., Svensk. Fanerogamfl. 101. 1918. Quito, Ecuador, *Humboldt & Bonpland*; P (holotype); US, fragment.

Festuca unioloides Willd., Hort. Berol. 3:pl. 3. 1803. *Ceratochloa unioloides* Beauv., Ess. Agrost. 75:pl. 15. f. 7. 1812. *Bromus unioloides* (Willd.) Raspail, Ann. Sci. Nat., Bot. 5:439. 1825. *B. willdenowii* Kunth, Rev. Gram. 1:134. 1829. *Tragus unioloides* Panz. ex Jacks., Ind. Kew. 2:1099. 1895. Type grown at Berlin from seed from Carolina where probably cultivated.

Ceratochloa haenkeana J. S. Presl *in* C. B. Presl, Reliq. Haenk. 1:285. 1830. *B. haenkeanus* (Presl) Kunth, Enum. Pl. 1:416. 1833. *B. unioloides* var. *haenkeanus* (Presl) Shear, U.S. Dept. Agr., Div. Agrost. Bull. 23:52. 1900. "In Cordilleris chilensibus inque montanis Peruviae."

B. strictus Brongn. in Duperrey, Voy. Autour Monde La Coquille, Bot., Phanerog. 2:45. 1831.

Ceratochloa pendula Schrad., Linnaea 6:Litt. 72. 1831. *B. schraderi* Kunth, Enum. Pl. 1:416. 1833. Type grown at Gottingen from seed from Carolina.

Ceratochloa breviaristata Hook. Fl. Bor. Am. 2:253. 1840. *B. breviaristatus* Thurb. in Wilkes, U.S. Expl. Exped. Bot. 17:493. 1874, *non* Buckl., 1862. *Forasaccus brebiaristatus* Lunell, Amer. Midl. Nat. 4:225, sphalm. 1915. "Lewis and Clark River and near the sources of the Columbia, *Douglas*, in 1826".

Ceratochloa submutica Steud. (?), Syn.Pl. Gram 321. 1854.

39. *B. subvelutinus* Shear, U.S. Dept. Agr., Div. Agrost. Bull. 23:52. 1900. Reno, Nevada, *Tracy* 249.

Bromus luzonensis Presl, Rel. Haenk. 1:262. 1830. Holotype: "Philippines, Luzon" (? California), *Haenke*.

40. *Bromus marginatus* Nees in Steud., Syn. Pl. Glum. 1:322. 1854. *B. hookeri* var. *marginatus* (Nees) Fourn., Mex. Pl. 2:127. 1886. *Ceratochloa*

marginata (Nees) Jacks., Ind. Kew. 1:487. 1893. *Forasaccus marginatus* (Nees) Lunell, Amer. Midl. Nat. 4:225. 1915. *B. sitchensis* var. *marginatus* (Nees) B. Boivin, Naturaliste Canad. 94:521. 1967. Columbia River, *Douglas*.

Bromus breviaristatus Buckl., Acad. Nat. Sci. Phila. Proc. 1862:98. 1862. *B. parviflorus* Nutt. ex A. Gray, Acad. Nat. Sci. Phila. Proc. 1862:336. 1862. *B. pauciflorus* Nutt. ex Shear, U.S. Dept. Agr., Div. Agrost. Bull. 23:53. 1900. Rocky Mountains, *Nuttall*. *B. marginatus* Nees var. *breviaristatus* (Buckl.) Beetle, Phytologia 55:209. 1984. As pointed out by Shear (1900), the type specimen of *B. breviaristatus* Buckl. belongs to *B. marginatus* Nees *sensu stricto*. Although making this new combination, Beetle (1984) applied it to specimens that are here taken to belong to *B. subvelutinus* Shear.

Bromus marginatus var. *seminudus* Shear, U.S. Dept. Agr., Div. Agrost. Bull. 23:55. 1900. Wallowa Lake, Oregon, *Shear 1811*.

Bromus marginatus var. *latior* Shear, U.S. Dept. Agr., Div. Agrost. Bull. 23:55. 1900. *B. latior* (Shear) Rydb., Fl. Rocky Mount. 89. 1917. Walla Walla, Washington, *Shear 1615*.

Bromus flodmanii Rydb., Torrey Bot. Club Bull. 36:538. 1909. Sheep Creek, Montana, *Flodman 187*.

41. *Bromus maritimus* (Piper) A. S. Hitchc. in Jeps., Fl. Calif. 1:177. 1912. *B. marginatus* subsp. *maritimus* Piper, Proc. Biol. Soc. Wash. 18:148. 1905. *B. carinatus* var. *maritimus* (Piper) C. L. Hitchc., Vasc. Pl. Pac. Northwest 1:504. 1969. *Ceratochloa maritima* (Piper) Holub, Folia Geobot. Phytotax. 8:170. 1973. Point Reyes, California, *Davy 6798*.

42. *Bromus polyanthus* Scribn. in Shear, U.S. Dept. Agr., Div. Agrost. Bull. 23:56. 1900. *B. multiflorus* Scribn. U.S. Dept. Agr., Div. Agrost. Bull. 13:46. 1898, *non B. multiflorus* Weigel, 1772. Battle Lake, Wyoming, *A. Nelson 4021*.

Bromus polyanthus var. *paniculatus* Shear, U.S. Dept. Agric., Div. Agrost. Bull. 23:57. 1900. *B. paniculatus* Rydb., Fl. Rocky Mount. 90. 1917. West Mancos Canyon, Colorado, *Tracy, Earle & Baker 333*.

? *B. laciniatus* Beal, Grasses N. Amer. 2:615. 1896. *Ceratochloa laciniata* (Beal) Holub, Folia Geobot. Phytotax. 8:170. 1973. Type: Mexico, Oaxaca, *Pringle 4897*; MSC (holotype); GH.

43. *Bromus sitchensis* Trin. in Bong., Mem. Acad. St. Petersb. 6, 2:173. 1832. Sitka, Alaska, *Mertens*.

44. *Bromus stamineus* Desv. in Gay, Hist. Chile Bot.6:440. 1856. Rancagua, Chile(?), *Bertero 117*.

Section *GENEA*

45. *Bromus diandrus* Roth, Bot. Abh. 44. 1787.
    B. rigidus var. *gussonei* (Parl.) Coss. & Dur., Expl. Sci. Alger. 2:159. 1855. *B. gussonei* Parl., Rar. Pl. Sic. 2:8. 1840. *B. maximus* var. *gussonei* (Parl.) Parl., Fl. Ital. 1:407. 1848. *B. villosus* var. *gussonei* (Parl.) Aschers. & Graebn., Syn. Mitteleur. Fl. 2:595. 1901. *Zerna gussonei* (Parl.) Grossh., Akad. Nauk U.S.S.R. Bot. Inst. Trudy Azerbaidzh. Fil. 8:305. 1939. Europe.

46. *Bromus madritensis* L., Cent. Pl. 1:5. 1755; Amoen. Acad. 4:265. 1759.
    Festuca madritensis (L.) Desf., Fl. Atlant. 1:91. *Zerna madritensis* (L.) Panz. ex Jacks., Ind. Kew. 2:1249 *Anisantha madritensis* (L.) Nevski, Act. Univ. Asiae Med. 8b. Bot. 17:21. 1934. Spain.

47. *Bromus rigidus* Roth, Bot. Mag. (Zurich) 4(10):21. 1790. *B. rubens* var. *rigidus* (Roth) Mutel, Fl. Franc. 4:133. 1837. *B. madritensis* var. *rigidus* (Roth) Bab. ex Syme in Sowerby, English Bot. ed 3, 11:161. 1873. *B. villosus* var. *rigidus* (Roth) Aschers. & Graebn., Syn. Mitteleur. Fl. 2:596. 1901. *Anisantha rigida* (Roth) Hylander, Uppsala Univ. Arskr. 7:32. 1945. Type: Unknown (Europe).
    B. villosus Forsk., Fl. Aegypt. Arab. 23. 1775; *non B. villosus* Scop., 1772. Egypt.
    B. maximus Desf., Fl. Atlant. 1:95. pl.26. 1798; *non* Gilib., 1790. *B. madritensis* var. *maximus* (Desv.) St. Amans, Fl. Agen. 45. 1821. *B. villosus* var. *maximus* (Desf.) Aschers. & Graebn., Syn. Mitteleur. Fl.2:595. 1901. *Forasaccus maximus* (Desf.) Bubani, Fl. Pyr. 4:382. 1901. North Africa.

48. *Bromus rubens* L., Cent. Pl. 1:5. 1755; Amoen. Acad. 4:265. 1759. *Festuca rubens* (L.) Pers., Syn. Pl. 1:94. 1805. *B. scoparius* var. *rubens* (L.) St. Amans, Fl. Agen. 45. 1821. *B. madritensis* subsp. *rubens* (L.) Husnot, Gram. Fr. Belg. 71. 1899. *Anisantha rubens* Nevski, Act. Univ. Asiae Med. 8b. Bot. 17:19. 1934. *Zerna rubens* Grossh., Akad. Nauk U.S.S.R. Bot. Inst. Trudy Azerbaidzh. Fil. 8:306. 1939. Type: Spain, *Loefling* (LINN 93.28).

49. *Bromus sterilis* L., Sp. Pl. 77. 1753. *Schedonorus sterilis* (L.) Fries, Bot. Not. 131. 1843. *Anisantha sterilis* Nevski, Act. Univ. Asiae Med. 8b. Bot. 17:20. 1934. Europe.
    Zerna sterilis Panz. ex Jacks., Ind. Kew. 2:1249. 1895; synonym of *B. sterilis* L.

50. *Bromus tectorum* L., Sp. Pl. 77. 1753. *Schedonorus tectorum* (L.) Fries, Bot. Not. 131. 1843. *Anisantha tectorum* (L.) Nevski, Act. Univ. Asiae Med. 8b. Bot. 17:20, 22. 1934. Europe.
B. setaceus Buckl., Acad. Nat. Sci. Phila. Proc. 1862:98. 1862. Northern Texas, *Buckley.*
Zerna tectorum Panz. ex Jacks., Ind. Kew. 2:1249. 1895; as synonym of *B. tectorum* L.
B. tectorum var. *glabratus* Spenner, Fl. Friburg. 1:152. 1825. Germany.
B. tectorum var. *nudus* Klett. & Richt., Fl. Leipzig 109. 1830. *B. tectorum* f. *nudus* (Klett. & Richt.) St. John, Fl. Southeast Wash. and adj. Idaho 36. 1937. Germany.

Section *NEOBROMUS*

51. *Bromus berterianus* Colla, Herb. Pedem. 6:68. 1836. Mem. Reale Accad. Sci. Torino 39:25. Pl. 58. 1836. Rancagua, Chili, *Bertero 117*; TO (holotype), P (isotype).
Bromus trinii Desv. in Gay, Fl. Chil. 6:441. 1853. *Trisetum hirtum* Trin., Linnaea 10:300. 1836; *non Bromus hirtus* Lichtst. *in* Roem. & Schult. 1817. *Trisetum trinii* (Desv.) Louis-Marie, Rhodora 30:243. 1928. *Trisetobromus hirtus* (Trin.) Nevski, Acta Univ. Asiae Med. 8b. Bot. 17:15. 1934. Chile; LE (syntype); US, fragment.
B. trinii var. *pallidiflorus* Desv. in Gay, Fl. Chil. 6:441. 1853. *Trisetum trinii* var. *pallidiflorus* (Desv.) Louis-Marie, Rhodora 30:243. 1928. Chile.
Trisetum barbatum Steud., Syn. Pl. Glum. 1:229. 1854; *non T. barbatum* Nees, 1841. *B. barbatoides* Beal, Grasses N. Amer. 2:614. 1896. Chile, *Bertero 806.*
Danthonia pseudo-spicata C. Muell., Bot. Ztg. 14:348. 1856. Valparaiso, Chile, *Cuming 466.*
Trisetum barbatum var. *major* Vasey, U.S. Dept. Agr., Div. Bot. Bull. 13:60. 1893. *T. trinii* var. *majus* Louis-Marie, Rhodora 30:243. 1928. Holotype: Mexico, *Palmer 667*; US.
B. barbatoides var. *sulcatus* Beal, Grasses N. Amer. 2:615. 1896. Mexico, *Palmer 667.*
B. trinii var. *excelsus* Shear, U.S. Dept. Agr., Div. Agrost. Bull. 23:25. 1900. *B. berterianus* var. *excelsus* (Shear) Pavlick (new combination made above). Panamint Mountains, California, *Coville & Funston 522.*

GLOSSARY

acuminate: gradually tapering to a narrow tip or sharp point; otherwise used in the sense of gradually and concavely tapering.
acute: sharp-pointed but less tapering than acuminate.
adventive: non-native plant introduced to an area and locally established or not naturalized.
alpine: those parts of mountains that rise above the cold limits of trees.
antrorse: directed upward or forward.
apical: pertaining to the apex or tip of a plant organ.
apices: plural of apex.
appressed: pressed flat against or close to another organ, as panicle branches against the axis or hairs against a stem or leaf.
arctic: pertaining to the unforested regions that occur north of the tree line.
arcuate: arching or moderately curved like a bow.
aristate: bearing an awn on the tip.
ascending: sloping upward, as culms which curve upward from the base; or branches of an inflorescence which slope upward at an angle of about 40°-70°.
asperity: a minute, sharp projection or prickle-hair of an epidermal cell.
auricles: projecting lobes or appendages, usually paired, and arising from the summit of the leaf sheath on either side of the throat.
auriculate: bearing auricles.
awn: a bristle, usually terminal to a lemma.
axis: general term for a central, supporting grass organ (e.g., a rachilla), or in particular, the main or central stem of a compound inflorescence (e.g., that of a panicle); compare *rachis*.
basal: that part near to or forming part of the base.
bifid: two-lobed at the apex.
bilobed: two-lobed.
bract: any of the reduced or modified leaves of the inflorescence and upper part of a shoot; in grasses, the glumes and lemmas are bracts.
caespitose: growing in tufts i.e. with several or many stems growing closely together.
callus-scar: scar left on the thickened base of the lemma after abscission from the rachilla.
canescent: hoary due to dense, fine, whitish or grayish hairs.
carinate: keeled.
caryopsis: the fruit of grasses; the grain.
chaparral: a vegetation type having a more or less closed cover of mostly hard-wooded and broad-leaved evergreen sclerophyll type shrubs restricted to areas of hot summers with

deficient precipitation, and subject to fires following which many of the shrubs stump-sprout.

chartaceous: papery in texture, often indicated by wrinkles.

chasmogamous: flowers or having flowers which open and expose their stamens and stigmas to the air; in grasses it is the floret which opens.

ciliate: with a fringe of hairs on the margin.

ciliolate: minutely ciliate.

cleistogamous: flowers or having flowers which are fertilized and set seed without opening; in grasses it is the floret which does not open.

collar: junction area of the leaf sheath and blade.

compressed: unusually flattened in one plane as the spikelets of some grasses.

coriaceous: leathery in texture, firmer than chartaceous.

corneous: of a texture resembling horn; in grasses, used to describe lemmas that are hard, firm or indurate.

contracted: closely bunched together as the panicle branches with the axis in some grasses.

culm: the above-ground stem of a grass.

cuneate: wedge-shaped, with the narrow end proximal.

cytogenetic: of the science of cytogenetics which combines cytological and genetical techniques in plant research.

decumbent: with a horizontal or inclined base which curves upward into an erect or ascending tip; as some grass culms.

depauperate: small or poorly developed; the condition of impoverished or dwarfed plants which are below average size.

distal: away or farthest from the point of attachment.

divaricate: widely and stiffly divergent from the axis of a plant organ; e.g., from a lemma, rachis or panicle axis.

drooping: bending downward, especially as the tips of panicle branches weighted down by spikelets.

endemic: confined to a particular, often relatively small, geographic area.

erect: standing upright, as the tip of some grass panicles.

erose: having an irregular margin, as if gnawed.

filiform: threadlike.

first glume: the lower glume; i.e., the proximal one.

flexuous: having relatively firm bends or more or less firmly serpentine.

floret: the basic unit of a grass spikelet, having one flower subtended and usually enclosed by the lemma and the palea.

geniculate: abruptly bent or twisted.

glabrous: without hairs.

glaucous: having a whitish or bluish cast due to surface texture or bloom.

glume: one of a pair of bracts, which do not subtend flowers, found at the base of a grass spikelet.

grass bald: an open area in an otherwise forested zone, and often in a mountainous region, in which grasses are a significant or dominant part of the vegetation; grass balds often occur in regions having summers with deficient precipitation, and are usually confined to areas of shallow soil (as on mountain summits occurring below the subalpine zone) and/or steep south-facing or west-facing slopes.

hairy: bearing hairs (trichomes) of any sort.

hirsute: studded with coarse or stiff, often bent or curved hairs.

hispidulous: having short, coarse and firm hairs.

hirtellous: minutely hirsute.
hyaline: thin and translucent or transparent.
indurate: hardened; firm; retaining its shape.
inflorescence: the flowering part of a plant; in grasses, the panicle, raceme, spike, etc.
innovation: the basal shoot of a perennial grass.
internerves: the area between the nerves of a leaf, lemma or other organ.
internode: the part of a stem or culm between two successive nodes.
involute: rolled inwards, as the margins of a grass leaf blade, exposing the lower surface and concealing or partially concealing the upper surface.
keeled: having a conspicuous, central, longitudinal ridge (resembling the keel of a ship), as the lemma of some grasses; carinate.
lacerate: torn, or with an irregularly jagged margin.
laciniate: cut or torn into narrow and often unequal segments.
lacustrine: pertaining to lakes.
lanate: having woolly hairs.
lanceolate: lance-shaped, long and tapered to a point distally.
lemma: the outer of the pair of bracts that subtend a grass flower.
ligule: in grasses applied to the tongue-shaped appendage arising at the junction of the leaf sheath and blade, and partially surrounding the culm.
macrohair: in grasses, unicellular, usually thick-walled hairs of the epidermis; short macrohairs are termed prickle-hairs.
montane: pertaining to mountain slopes, and often, in particular, to the mostly forested zone extending downslope from the subalpine forest zone.
mucronate: tipped with a mucro, a short, sharp slender point; e.g., a short awn.
nerve: a longitudinal vein of a leaf, lemma or other organ.
nodding: bent to the side, as is the tip of a grass panicle that is not erect.
node: the part of a stem to which a leaf is attached; in grasses this is often a thickened part of the culm.
obcompressed: flattened opposite to the usual way (e.g. flattened dorso), ventrally instead of laterally.
obovate: shaped as the long section through a hen's egg, and with the narrow end being proximal.
obovoid: egg-shaped, with the narrow end being proximal.
octoploid: having eight genomes i.e. eight basic sets of chromosomes.
ovate: shaped as the long section through a hen's egg, and with the narrow end being distal.
ovoid: egg-shaped, with the narrow end being distal.
palea: the inner of the pair of bracts that subtend a grass flower.
panicle: a compoundly branched inflorescence, as in many grasses.
patent: turned outward at an angle of more or less 90°, as the angle some panicle branches make with their axis.
pedicel: the stalk of a single flower; in grasses, the stalk of a single spikelet.
pilose: having long, more or less straight, soft, spreading hairs.
polyploid: having three or more genomes, i.e. three or more basic sets of chromosomes; any ploidy level above the diploid.
prickle-hairs: short macrohairs; asperities.
prow-shaped: shaped as the forward part of a ship; said of some grass leaf-blade tips that have their margins turned upward—and sometimes inwards as well— resembling a ship's prow.
proximal: near or nearest to the point of attachment.

puberulent: minutely pubescent i.e. having very short hairs.
pubescent: having short, soft hairs; i.e., hairy; sometimes loosely used for bearing hairs (trichomes) of any sort.
racemose: in grasses, referring to an inflorescence having simple panicle branches (pedicels) on a rachis which bear one spikelet each.
rachilla: the axis of a spikelet.
rachis: in grasses, the main or central stem of a simple (non-compound) inflorescence such as that of a raceme or spike (compare *axis*).
recurved: curved backward.
retrorse: directed downward or backward.
rhizomatous: having rhizomes.
rhizomes: the underground stems of a plant; e.g., that of some grasses.
scabridulous: minutely scabrous.
scabrous: rough to the touch, usually because of the presence of minute asperities or prickle-hairs on the epidermis.
scarious: thin, dry, membranaceous and not green.
second glume: the upper glume; i.e., the distal one.
secund: in grasses, a panicle or raceme having all the branches turned to one side of the axis or rachis, usually by twisting of the branches or pedicels.
sericeous: silky, from having long, slender, soft, more or less appressed hairs.
sheath: the basal part of a grass leaf which surrounds the culm.
sinuous: very slender (fine), flexible and curving one way then another, sometimes irregularly.
spikelet: in grasses, the compact unit of the inflorescence consisting of the (usually) two glumes and one to several florets borne on the rachilla.
spreading: directed outward; in grasses, often used to refer to hairs on an organ (e.g. a leaf sheath) or to panicle branches which form an angle with the axis between that of ascending and patent; i.e., between about 70°-90°.
striate: having fine parallel lines or minute ridges.
subapical: below the apex.
subalpine: pertaining to the forest zone immediately below the alpine tundra.
submontane: pertaining to the steppe zone in mountainous areas occurring below the montane zone.
subracemose: not quite racemose; in grasses, having most spikelets borne singly on simple branches (pedicels) arising from a rachis, but with one or very few branches compound and bearing two spikelets.
terete: cylindrical or rounded in cross-section.
throat: variously the area between or the angle formed by the upper margins of the leaf sheath.
truncate: as if cut off; having the apex or base of an organ transversely straight or nearly straight.
verticillate: in grasses, having more or less dense whorls of panicle branches.
villous: having long, soft, often bent or curved hairs.

BIBLIOGRAPHY

Albee, B. J., L. M. Shultz and S. Goodrich. 1988. Atlas of the vascular plants of Utah. Occasional Publication No. 7, Utah Museum of Natural History, Salt Lake City.
Amman, K. 1981. Bestimmungsschwierigkeiten bei europäischen *Bromus*-Arten. Bot. Jahrb. Syst. 102: 459-469.
Anderson, J. P. 1944. Flora of Alaska and adjacent parts of Canada. Part 2. *Typhaceae* to *Poaceae*. Iowa State J. Sci. 18:381-445.
Anstey, T.H. 1986. One Hundred Harvests. Research Branch, Agriculture Canada Historical Series No. 27. Ottawa.
Armstrong, K.C. 1981. The evolution of *Bromus inermis* and related species of *Bromus* sect. *Pnigma*. Bot. Jahrb. Syst. 102:427-443.
_____. 1983. The relationship between some Eurasian and American species of *Bromus* sect. *Pnigma* as determined by the Karyotypes of some F-1 hybrids. Can J. Bot. 61:700-707.
_____. 1984. Chromosome pairing affinities between old-world and new-world species of *Bromus* sect. *Pnigma*. Can. J. Bot. 62:581-585.
Baum, B.R. 1967. Kalm's specimens of North American grasses; their evaluation for typification. Can. J. Bot. 45:1845-1852.
Beal, W.J. 1896. Grasses of North America for farmers and students. Vol. 2. Henry Holt & Co., New York.
Beetle, A.A. 1984. Nomenclatorial changes involving Wyoming grasses. Phytologia 55:209-213.
Böcher, T.W., K. Holmen and K.Jakobsen. 1968. The flora of Greenland. P. Haase & Son, Publishers, Copenhagen.
Boivin, B. 1981. Flora of the Prairie provinces. Part V. *Gramineae*. Provancheria 12. Mémoires de l'Herbier Louis-Marie. Université Laval, Québec.
Bongard, M. 1832. Observations sur la végétation de l'Ile de Sitcha. Mém. Acad. Impér. Sci. St. Petersb. 6 Sér. 2:173.
Bowden, W.M. 1960. Chromosome numbers and taxonomic notes on northern grasses. 2. Tribe Festuceae. Can. J. Bot. 38:117-131.
Breitung, A.J. 1957. Annotated catalogue of the vascular flora of Saskatchewan. Am. Midl. Nat. 58:1-72.
Calder, J.A. and R.L. Taylor. 1968. Flora of the Queen Charlotte Islands. Part 1. Canada Dept. Agric. Monogr. No. 4. Ottawa.
Clapham, A.R., T.G. Tutin and E.F. Warburg. 1965. Flora of the British Isles, illustrations. Part 4. *Monocotyledones*. Cambridge University Press, Cambridge.

Clayton, W.D. and S.A. Renvoize. 1986. Genera graminum. Grasses of the world. Kew Bulletin Additional Series XIII. Royal Botanic Gardens, Kew. 389 pp.

Correll, D.S. and M.C. Johnston. 1970. Manual of the vascular plants of Texas. Texas Research Foundation, Renner. 1881 pp.

Coulter, J.M. 1885. Manual of the Botany of the Rocky Mountain region from New Mexico to the British Boundary. Ivison, Blakeman, Taylor and Co., New York.

Cugnac, A. and A. Camus. Sur quelques bromes et leurs hybrides. Bull. Soc. Bot. Fr. I-IV, 78:327-341. 1931; 80:561-562. 1933; 81:318-323. 1934; 83:47-68. 1936; 83:658-667. 1936; 84:437-440. 1937; 84:711-713. 1937.

Daubenmire, R. Plant geography with special reference to North America. 1978. Academic Press, New York. 338 pp.

Davis, P.H. and V.H. Heywood. 1973. Principles of angiosperm taxonomy. Robert E. Krieger Publishing Co., Huntington, New York.

Desvaux, E. 1853. Gramineas. In C. Gay. Historia fisica y politica de Chile. Botanica. 6. El Museo de Historia Natural de Santiago, Chile.

Dore, W.G. and J. McNeill. 1980. Grasses of Ontario. Monograph 26. Research Branch, Agriculture Canada, Ottawa.

Dorn, R.D. 1977. Manual of the vascular plants of Wyoming. Garland Publishing, Inc., New York. 1498 pp.

Elliot, F.C. 1949. *Bromus inermis* and *B. pumpellianus* in North America. Evolution 3:142-149.

Eastham, J.W. 1947. Supplement to 'flora of southern British Columbia' (J.K. Henry). Special Publication No. 1, British Columbia Provincial Museum. Victoria, B.C.

Ekman, J. 1989. Sloklosta *Bromus sitchensis* och plattlosta *B. willdenowii* i Sverige. (*Bromus sitchensis* and *B. wildenowii* in Sweden.) Svensk. Bot. Tidskr. 83:87-100..0

Fernald, M.L. 1930. The complex *Bromus ciliatus*. Rhodora 32:63-71.

_____. 1941. Another century of additions to the flora of Virginia. Rhodora 43:485-553;559-657.

Gill, G.S. and S.A. Carstairs. 1988. Morphological, cytological and ecological discrimination of *Bromus rigidus* from *Bromus diandrus*. Weed Research 28:399-405.

Gleason, H.A. 1958. The new Britton and Brown illustrated flora of the northeastern United States and adjacent Canada. Vol. 1. The *Pteridophyta, Gymnospermae* and *Monocotyledoneae*. The New York Botanical Garden, New York.

Gleason, H.A. and A. Cronquist. 1963. Manual of the vascular plants of the northeastern United States and adjacent Canada. Van Nostrand Reinhold Co., New York. 810 pp.

_____. 1991. Manual of vascular plants of northeastern United States and adjacent Canada. Second Edition. The New York Botanical Garden, Bronx.

Gould, F.W. 1951. Grasses of the southwestern United States. Univ. Ariz. Biol. Sci. Bull. 7:1-352.

Gould, F.W. and R. Moran. 1981. The grasses of Baja California, Mexico. San Diego Society of Natural History, San Diego. 140 pp.

Gray, A. 1848. A manual of botany of the northern United States. 710 pp.

_____. 1859. Manual of the botany of the northern United States. Rev. ed. 739 pp.

Hansen, A.A. 1972. Grass varieties in the United States. US Dept. Agric. Handbook No. 170:1-124.

Harlan, J.R. 1945a. Cleistogamy and chasmogamy in *Bromus carinatus* Hook. & Arn. Amer. J. Bot. 32:66-72.

_____. 1945b. Natural breeding structure in the *Bromus carinatus* complex as determined by population analyses. Amer. J. Bot. 32:142-147.

Harrington, H.D. 1954. Manual of the plants of Colorado. Sage Books, Denver. 666 pp.

Henry, J.K. 1915. Flora of southern British Columbia and Vancouver Island. W.J. Gage & Co. Ltd., Toronto.

Hill, H.D. and W.M. Myers. 1948. Chromosome number in *Bromus inermis* Leyss. Jour. Amer. Soc. Agron. 40:466-469.

Hitchcock, A.S. 1934. New species and changes in nomenclature, of grasses of the United States. Amer. J. Bot. 21:127-139.

_____. 1935. Manual of the grasses of the United States. U.S. Dept. Agric. Misc. Publ. 200:1-1040.

Hitchcock, A.S. and A. Chase. 1951. Manual of the grasses of the United States. 2nd ed. US Dept. Agric. Misc. Publ. 200:1-1051.

Hitchcock, C.L. 1969. *Gramineae*. In C.L. Hitchcock, A. Cronquist, and M. Ownbey. Vascular plants of the Pacific Northwest. Part 1. Vascular cryptogams, gymnosperms and monocotyledons. University of Washington Press, Seattle.

Holmberg, O.R. 1924. *Bromi molles*, eine nomenklatorische und systematische Untersuchung. Bot. Not. 1924:313-328.

_____. 1926. Ueber die Begrenzung und Einteilung der Gramineen-Tribus *Festuceae* und *Hordeae*. Bot. Not. 1926:69-80.

Holmgren, A.H. & N.H. Holmgren. 1977. *Poaceae*. Pp. 184-464. In A. Cronquist, A.H. Holmgren, N.H. Holmgren, J.A. Reveal and P. Holmgren (eds.). Intermountain flora. Vascular plants of the intermountain west, U.S.A. Vol. 6. The Monocotyledons. Columbia University Press, New York.

Holub, J. 1973. New names in *Phanerogamae* 2. Folia Geobot. Phytotax., Praha, 8:155-179.

Hooker, W.J. 1840. Flora Boreali-Americana. Vol. 2. Henry G. Bohn, London.

Hooker, W.J. and G.A.W. Arnott. 1841. The Botany of Captain Beechey's voyage; comprising an account of the plants collected by Messrs Lay and Collie, and other officers of the expedition, during the voyage to the Pacific and Bering's Strait, performed in His Majesty's Ship Blossom. Henry G. Bohn, London.

Hubbard, C.E. 1934. *Gramineae*. In J. Hutchinson, Families of flowering plants 2:199-229.

_____. 1948. *Gramineae*. In J. Hutchinson. British flowering plants, 284-348.

_____. 1956. *Bromus catharticus* Vahl. (Symb. Bot. 2:22, 1791) versus *Bromus unioloides* H.B.K. (Nov. Gen. et Sp.,1:151, 1816) versus *Bromus unioloides* (Willd.) Rasp. (Ann. Sci. Nat., Bot. 5:439, 1825). Agron. Lusit. 18:7.

_____. 1968. Grasses. 2nd ed. Penguin Books Ltd., Harmondsworth, Middlesex, England.

Hubbard, W.A. 1955. The grasses of British Columbia. Handbook No. 9. British Columbia Provincial Museum, Victoria.

Hudson, J.H. 1985. Taxonomic reminder for recognizing Saskatchewan plants. Saskatchewan Natural History Society, Saskatoon.

Hultén, E. 1937. Flora of the Aleutian Islands. Vol. 1. Stockholm; 1960, 2nd ed., J. Cramer, Weinheim / Berstr.

_____. 1942. Flora of Alaska and Yukon. Vol. 2. Monocotyledoneae. Lunds Univ. Arsskr. Avd. 2, 38:129-412.

_____. 1964. The circumpolar plants. Vascular cryptogams, conifers, monocotyledons. Kungl. Svensk. Vetenskapsakad. Handl., ser. 4, 8:1-280.

_____. 1968. Flora of Alaska and neighboring territories. Stanford University Press, Stanford.

Hylander, N. 1945. Nomenklatorische und Systematische studien über Nordische Gefässpflanzen. Uppsala Univ. Årsskr. 1(7):1-337.

Jepson, W.L. 1963. A manual of the flowering plants of California. University of California Press, Berkeley and Los Angeles. 1238 pp.

Kartesz, J.T. 1994. A synonymized checklist of the vascular flora of the United States, Canada and Greenland. Second Edition. Vol. 2. Thesaurus. Timber Press, Portland, Oregon. 816 pp.

Kartesz, J.T. and R. Kartesz. 1980. A synonymized checklist of the vascular flora of the United states, Canada and Greenland. Vol. 2. The biota of North America. The University of North Carolina Press, Chapel Hill.

Kerguélen, M., 1975. Les *Gramineae* (*Poaceae*) de la flore française. Essai de mise au point taxonomique et nomenclaturale. Lejeunia, nouv. sér., 75:1-343.

_____, 1981. (Corrections et commentaires..) 8913(17). *Bromus molliformis*. Bull. Soc. Ech.Pl. Vasc. Eur. Bass. Medit. (Liège), 18:27.

_____, 1983. Les Graminées de France au travers de "Flora Europaea" et de la "Flore" du CNRS. Lejeunia, nouv. ser., 110:1-79.

_____, 1987. Données taxonomiques, nomenclaturales et chorologiques pour une révision de la flore de France. Lejeunia, nouv. sér. 120:1-264.

Knowles, P. F. 1944. Interspecific hybridizations of *Bromus*. Genetics 29:128-140.

Knowles, R.P., V.S. Baron and D.H. McCartney. 1993. Meadow bromegrass. Agriculture Canada Publication 1889/E. Agriculture Canada, Ottawa.

Lamson-Scribner, F. & E.D. Merrill. 1910. The grasses of Alaska. Contr. US Natl. Herb. 13:47-92.

Ledebour, K.F. von. 1829. Flora altaica 1:1-440.

_____. 1853. Flora rossica 4:1-741.

Lindman, C. A. M. 1926. Svensk fanerogamflora, ed. 2. 644 pp.

Linnaeus, C. 1753. Species plantarum. 1200 pp.

Looman, J. and K.F. Best. 1979. Budd's flora of the Canadian Prairie Provinces. Agriculture Canada Publication No. 1662, Ottawa.

Maire, R. 1941. In L. Emberger and R. Maire, Catalogue des Plantes du Maroc, 4. p. 943.

Maire, R. and M. Weiller. 1955. Flore de l'Afrique du Nord. Vol. 3. Monocotyledonae. Paul Lechevalier, Paris.

Marie-Victorin, Frère. 1964. Flore Laurentienne. 2nd. ed. Revised by E. Rouleau. Les Presses de l'Université de Montréal, Montréal.

Matthei, O. 1986. El genero *Bromus* L. (Poaceae) en Chile. Gayana, Bot. 43(1-4):47-110.

McNeill, J. 1976. Nomenclature of four perennial species of *Bromus* in eastern North America, with a proposal for the listing of *B. purgans* L. as a rejected name under Article 69. Taxon 25:611-616.

_____. 1977. *Bromus latiglumis* (Shear) A. S. Hitchc. need not be replaced by *B. altissimus* Pursh. Taxon 26:584.

Mitchell, W.W. 1965. Redefinition of *Bromus ciliatus* and *B. richardsonii*. Brittonia 17:278-284.

_____. 1967. Taxonomic synopsis of *Bromus* section *Bromopsis* (*Gramineae*) in Alaska. Can. J. Bot. 45:1309-1313.

Mitchell, W.W. and A.C. Wilton. 1966. A new tetraploid brome, section *Bromopsis*, of Alaska. Brittonia 18:162-166.

Moore, D.M. 1983. Flora of Tierra del Fuego. Missouri Botanical Garden, St. Louis.

Morrow, L.A. and P.W. Stahlman. 1984. The history and distribution of downy brome (*Bromus tectorum*) in North America. Weed Science 32, Supplement 1:2-6.

Morton, J.K. and J.M. Venn. 1984. The flora of Manitoulin Island and the adjacent islands of Lake Huron, Georgian Bay and the North Channel. 2nd revised ed. University of Waterloo, Waterloo.

Moss, E.H. 1983. Flora of Alberta. 2nd ed. Revised by J. G. Packer. University of Toronto Press, Toronto.

Munz, P. A. and D.D. Keck. 1963. A California flora. University of California Press, Berkeley and Los Angeles.

Munz, P.A. 1968. Supplement to "A California flora". University of California Press, Berkeley and Los Angeles.
Nevski, S.A. 1934. Bromae tribus graminearum naturalis. Acta Univ. Asiae Mediae 8b. Bot. 17:14, 15.
Nevski, S.A. & V. B. Sochava. 1934. *Bromus* subg. *Zerna*. In V.L. Komarov. Flora USSR 2:554-568.
Nielsen, E.L. & L.M. Humphrey. 1937. Grass studies of chromosome numbers in certain members of the tribes *Festuceae, Hordeae, Phalarideae* and *Tripsaceae*. Amer. J. Bot. 24:276-279.
Nicora, E.G. 1978. Flora Patagonica, Part 3. *Gramineae*. Colleccion Cientifica del INTA., Buenos Aires, Argentina.
Parodi, L.R. 1947. Las gramineas del genero *Bromus* adventicias en la Argentina. Revista Arget. Agron. 14:1-19.
_____. 1956. Noticia sobre el ejemplar tipo de. "*Bromus catharticus*" Vahl. Rev. Argent. Agron. 23:115-121.
Pavlick, L.E. 1982. The *Festuca* and *Vulpia* of British Columbia: enumeration, descriptions, distributions, taxonomy. 58 pp. Unpublished report filed in the Royal British Columbia Museum library, Victoria, B.C.
_____. 1983a. The taxonomy and distribution of *Festuca idahoensis* in British Columbia and northwestern Washington. Can. J. Bot. 61:345-353.
_____. 1983b. Notes on the taxonomy and nomenclature of *Festuca occidentalis* and *F. idahoensis*. Can. J. Bot. 61:337-344.
_____. 1983c. *Festuca viridula* (*Poaceae*): Re-establishment of its original lectotype. Taxon 32:117-120.
_____. 1984. Studies on the *Festuca ovina* complex in the Canadian Cordillera. Can. J. Bot. 62:2448-2462.
_____. 1985. A new taxonomic survey of the *Festuca rubra* complex in northwestern North America, with emphasis on British Columbia. Phytologia 57:1-17.
Pavlick, L.E. and J. Looman 1984. Taxonomy and nomenclature of rough fescues, *Festuca altaica, F. campestris* (*F. scabrella* var. *major*), and *F. hallii* in Canada and the adjacent part of the United States. Can. J. Bot. 62:1739-1749.
Perring, F.H. 1962. *Bromus interruptus* (Hack.) Druce a botanical dodo? Nature in Cambridgeshire 5:28-30.
Pinto-Escobar, P. 1976. Nota sobre el ejemplar tipo de "*Bromus catharticus*" Vahl. Caldasia 11:9-16.
_____. 1981. The genus *Bromus* in northern South America. Bot. Jahrb. Syst. 102:445-457.
Piper, C.V. 1905. Proc. Biol. Soc. Wash. 18:148.
_____. 1906. Flora of the state of Washington. Contrib. US Natl. Herb.11:1-637.
Polunin, N. 1940. Botany of the Canadian eastern Arctic. Part 1. *Pteridophyta* and *Spermatophyta*. Natl. Mus. Can. Bull.92:1-408.
_____. 1959. Circumpolar Arctic flora. Claredon Press, Oxford.
Porsild, A.E. 1955. The vascualr plants of the western Canadian Arctic Archipelago. Natl. Mus. Can. Bull. 135:1-226.
_____. 1957. Illustrated flora of the Canadian Arctic Archipelago. Natl. Mus. Can. Bull. 146:1-209.
Porsild, A.E. and W.J. Cody. 1980. Vascular plants of the continental Northwest Territories, Canada. National Museums of Canada, Ottawa.
Presl, C.B. 1830. Reliquiae Haenkeanae. J.G. Calve, Bibliopolam, Prague.
Raven, P.H. 1960. The correct name for rescue grass. Brittonia 12:219-221.

Rydberg, P.A. 1917. Flora of the Rocky Mountains and adjacent plains. P.A. Rydberg, New York.
Sales, F. 1993. Taxonomy and nomenclature of *Bromus* sect. *Genea*. Edinb. J. Bot. 50:1-31.
Scholz, H. 1970. Zur Systematik der Gattung *Bromus* L. Subgenus *Bromus(Gramineae)*. Willdenowia 6:139-159.
Schulz-Schaeffer, J. 1956. Cytologische Untersuchungen in der Gattung *Bromus* L. Zeits. Pflanzenzücht. 35:297-320.
_____. 1960. Cytological investigations in the genus *Bromus*. 3. The cytotaxonomic significance of the satellite chromosomes. Jour. Hered. 51:269-277.
_____. and D. Markarian. 1957. Cytologische Untersuchungen in der Gattung *Bromus* L. Zeits. Pflanzenzücht. 37:299-316.
Scoggan, H. J. 1957. Flora of Manitoba. Natl. Mus. Can. Bull. 140:1-619.
_____. 1978. The flora of Canada. Part 2. National Museums of Canada, Ottawa.
Seymour, F. C. 1966. *Bromus mollis* and allies in New England. Rhodora 68:168-174.
Shear, C.L. 1900. A revision of the North American species of *Bromus* occurring north of Mexico. U.S. Dep. Agric. Bull. 23.
_____. 1901. Notes on Fournier's Mexican species and varieties of *Bromus*. Bull. Torrey Club 28:242-246.
Simpson, D.R. 1935. A study of species complexes in *Agrostis* and *Bromus*. Ph.D. Thesis. University of Washington, Seattle. 88 p.
Smith, P. 1965. Experimental taxonomy of *Bromus*. Ph.D. Thesis. University of Birmingham, Birmingham.
_____. 1968. The *Bromus mollis* aggregate in Britain. Watsonia 6:327-344.
_____. 1970. Taxonomy and nomenclature of the brome-grasses (*Bromus* L. s. l.). Notes Roy. Bot. Gard. Edinb. 30:361-375.
Smith, P.M. 1972. Serology and species relationships in annual bromes (*Bromus* L. sect. *Bromus*) Ann. Bot. 36(144):1-30.
_____. 1980. *Bromus* L. In Flora Europaea. Vol. 5. *Alismataceae* to *Orchidaceae* (*Monocotyledones*). Edited by T.G. Tutin, V.H. Heywood, N.A. Burges, D.M. Moore, D.H. Valentine, S.M. Walters and D.A. Webb. Cambridge University Press, Cambridge.
_____. 1981. Ecotypes and subspecies in annual brome grasses. *Bromus. Gramineae*. Bot. Jahrb. Syst. 102:497-510.
Soderstrom, T.R. and J.H. Beaman. 1968. The genus *Bromus* (*Gramineae*) in Mexico and Central America. Publications of the Museum, Michigan State University Biological Series 3:465-520.
Stapf, O. 1928. The nomenclature of *Bromus*. Kew. Bull. 1928:209-211.
Stebbins, G.L., Jr. 1947. The origin of the complex of *Bromus carinatus* and its phytogeographic implications. Contr. Gray Herb. 165:42-55.
_____. 1949. The evolutionary significance of natural and artificial polyploids in the family *Gramineae*. Proc. 8th Int. Cong. Genet. (Hereditas suppl. vol.) 461-485.
_____. 1981. Chromosomes and evolution in the genus *Bromus. Gramineae*. Bot. Jahrb. Syst. 102:359-380.
Stebbins, G.L., Jr. and R.M. Love. 1941. A cytological study of California forage grasses. Amer. J. Bot. 28:371-382.
Stebbins, G.L. and B. Crampton. 1960. A suggested revision of the grass genera of temperate North America. In Recent advances in botany from lectures and symposia presented to the IX International Botanical Congress, Montreal, 1959, 1:133-145. Toronto University Press, Toronto.
Stebbins, G.L., Jr. and H.A. Tobgy. 1944. The cytogenetics of hybrids in *Bromus*. 1. Hybrids within the section Ceratochloa. Amer. J. Bot. 31:1-11.

Stebbins, G.L., Jr. H.A. Tobgy and J.R. Harlan. 1944. The cytogenetics of hybrids in *Bromus*. 2. *Bromus carinatus* and *Bromus arizonicus*. Proc. Calif. Acad. Sci. 25:307-321.

Taylor, R.L. and B. MacBryde. 1977. Vascular plants of British Columbia. The Botanical Garden, The University of British Columbia. Tech. Bull. 4. The University of British Columbia Press, Vancouver.

Tidestrom, I. 1925. Flora of Utah and Nevada. Contr. U.S. Natl. Herb. 25:1-665.

Tidwell, W.D., S.R. Rushforth and D. Simper. 1972. Evolution of the floras in the Intermountain Region. In A. Cronquist, A.H. Holmgren, N.H. Holmgren and J. L. Reveal. 1972. Intermountain flora. Vol. 1. Hafner Publishing Co., Inc., New York.

Tournay, R. 1961. La nomenclature des sections du genre *Bromus* L. (*Gramineae*). Bull. Jard. Bot. Bruxelles 31:289-299.

_____. 1968. Le brome des Ardennes, «*Bromus arduennensis*», et ses proches, *B. secalinus* et *B. grossus*. Bull. Jard. Bot. Nat. Belg. 38:295-380.

Tsvelev, N.N. 1984. Grasses of the Soviet Union. Parts 1 & 2. Russian Translation Series 8. A. A. Balkema, Rotterdam. 1196 pp.

Upadhyaya, M.K., R. Turkington and D. McIlvride. The biology of Canadian weeds. 75. *Bromus tectorum* L. Can. J. Plant Sci. 66:689-709.

Vasey, G. 1885. Some new grasses. Bot. Gaz. 10:223-224.

_____. 1893. Grasses of the Pacific Slope, including Alaska and the adjacent islands. Part 2. U.S. Dep. Agric. Bull. 13.

Veldkamp, J.F. 1990. *Bromus luzonensis* Presl is the correct name for *Bromus breviaristatus* Buckl. (*Gramineae*). Taxon 39:660.

Vivant, J. 1964. Au sujet de *Bromus commutatus* Shrader. Bull. Soc. Bot. France 111:97-100.

Wagnon, H.K. 1950a. Three new species and one new form in *Bromus*. Leafl. West. Bot. 6:64-69.

_____. 1950b. Nomenclatural changes in *Bromus*. Rhodora 52:209-215.

_____. 1952. A revision of the genus *Bromus*, section *Bromopsis*, of North America. Brittonia 7:415-480.

Walters, M.S. 1966. Development and chemical constitution of a nuclear body in microsporocytes of *Bromus*. Heredity 21:173-181.

Weber, W.A. 1976. Rocky Mountain flora. Colorado Assoc. Univ. Press, Boulder. 479 pp.

Welsh, S. 1974. Anderson's flora of Alaska and adjacent parts of Canada. Brigham Young University Press, Provo.

Welsh, S.L., N.D. Atwood, S. Goodrich and L.C. Higgins (eds.) 1987. A Utah flora. Great Basin Naturalist Memoirs No. 9. Brigham Young University, Provo.

Wendelbo, P. 1956. Anthropochore *Bromus*-arten i Norge. Blyttia 14:1-3.

Wiegand, K.M. 1922. Notes on some east-American species of *Bromus*. Rhodora 24:89-92.

Wiggins, I.L. 1980. Flora of Baja California. Stanford University Press, Stanford.

Wilken, D.H. and E.L. Painter. 1993. *Bromus*. In J.C. Hickman (ed.) 1993. The Jepson Manual. Higher Plants of California. University of California Press, Berkeley.

INDEX

Anisantha madritensis 143
Anisantha rigida 143
Anisantha rubens 143
Anisantha sterilis 144
Anisantha tectorum 144
Avena secalinus 139
Brachypodium commutatum 137
Bromopsis anomala 131
Bromopsis canadensis 131
Bromopsis erectus 132
Bromopsis ciliata 131
Bromopsis frondosa 132
Bromopsis grandis 132
Bromopsis inermis 132
Bromopsis kalmii 133
Bromopsis laevipes 133
Bromopsis lanatipes 133
Bromopsis macroglumis 134
Bromopsis nottowayana 134
Bromopsis orcuttiana 134
Bromopsis pacifica 134
Bromopsis porteri 134
Bromopsis pseudolaevipes 135
Bromopsis pubescens 135
Bromopsis pumpelliana 135
Bromopsis richardsonii 136
Bromopsis suksdorfii 136
Bromopsis texensis 136
Bromopsis vulgaris 136
Bromus abolinii 139
Bromus aleutensis 7, 20, 94, 95, **96**, 106, 112, 140
Bromus alopecuros **130**
Bromus altissimus 133
Bromus altissimus f. incanus 133
Bromus anomalus 7, 14, 16, **24**, 130, 131
Bromus anomalus var. lanatipes 40, 133

Bromus arcticus 136
Bromus arenarius 18, **70**, 137, 138
Bromus arizonicus 19, 94, **98**, 140
Bromus arvensis 7, 17, 18, **72**, 137
Bromus arvensis var. patulus 138
Bromus arvensis var. racemosus 139
Bromus barbatoides 144
Bromus barbatoides var. sulcatus 144
Bromus berterianus 21, **128**, 144
Bromus berterianus var. excelsus **128**, 144
Bromus biebersteinii 8
Bromus brachyphyllus 134
Bromus breviaristatus 6, 104, 141, 142
Bromus brittanicus 139
Bromus brizaeformis 74, 137
Bromus briziformis 17, **74**, 137
Bromus californicus 140
Bromus canadensis 131
Bromus carinatus 6, 7, 19, 94, 95, **100**, 106, 130, 140
Bromus carinatus var. arizonicus 140
Bromus carinatus var. californicus **100**, 140
Bromus carinatus var. densus 141
Bromus carinatus var. hookerianus 19, **100**, 101, 140
Bromus carinatus var. linearis **100**, 140
Bromus carinatus var. maritimus 142
Bromus catharticus 8, 19, 94, **102**, 141
Bromus chiapporianus 139
Bromus ciliaris 131
Bromus ciliatus 8, 14, 16, **26**, 131, 133
Bromus ciliatus f. denudatus **26**, 132
Bromus ciliatus f. laeviglumis 135
Bromus ciliatus subvar. denudatus 132
Bromus ciliatus var. coloradensis 135
Bromus ciliatus var. denudatus 132
Bromus ciliatus var. glaberrimus 136

Bromus ciliatus var. *incanus* 133
Bromus ciliatus var. *incanus* subvar.
    latiglumis 133
Bromus ciliatus var. *intonsus* **26**, 132
Bromus ciliatus var. *laeviglumis* 135
Bromus ciliatus var. *latiglumis* 133
Bromus ciliatus var. *ligulatus* 136
Bromus ciliatus var. *minor* 131
Bromus ciliatus var. *montanis* 134
Bromus ciliatus var. *montanus* 134
Bromus ciliatus var. *pauciflorus* 136
Bromus ciliatus var. *porteri* 134
Bromus ciliatus var. *purgans* 135
Bromus ciliatus var. *purgans* subvar.
    laevivaginatus 135
Bromus ciliatus var. *scariosus* 134
Bromus commutatus 7, 8, 18, **76**, 86, 137
Bromus commutatus var. *apricorum* **76**, 137
Bromus commutatus subsp. *neglectus* 137
Bromus confertus 138
Bromus cyri 139
Bromus debilis 136
Bromus diandrus 8, 20, **116**, 143
Bromus divaricatus 138
Bromus dudleyi 132
Bromus erectus 8, 12, **28**, 132
Bromus erectus var. *arvensis* 137
Bromus eximius 136
Bromus eximius var. *umbraticus* 137
Bromus flodmanii 142
Bromus frondosus 7, 15, **30**, 132
Bromus gedrosianus 139
Bromus gracilis 139
Bromus gracilis var. *micromollis* 138
Bromus grandis 13, 14, 15, 16, **32**, 132
Bromus gussonei 143
Bromus haenkeanus 141
Bromus hookeri var. *ciliatus* 131
Bromus hookeri var. *canadensis* 131
Bromus hookeri var. *marginatus* 142
Bromus hookeri var. *pubescens* 135
Bromus hookerianus 140
Bromus hookerianus var. *minor* 141
Bromus hordeaceus 6, 18, **78**, 137
Bromus hordeaceus subsp. *divaricatus* **78**,
    80, 130, 138
Bromus hordeaceus subsp. *molliformis* 138
Bromus hordeaceus subsp. *mollis* 138
Bromus hordeaceus subsp. *pseudothominei*
    78, 80, 85, 86, 138

Bromus hordeaceus subsp. *thominei* **78**, 80
Bromus hordeaceus var. *mollis* 138
Bromus hordeaceus var. *pratensis* 137
Bromus hordeaceus var. *racemosus* 139
Bromus imperialis 133
Bromus incanus 133
Bromus inermis 8, 12, **34**, 132
Bromus inermis f. *aristatus* 132
Bromus inermis f. *bulbiferus* 132
Bromus inermis f. *villosus* 132
Bromus inermis subsp. *pumpellianus* 135
Bromus inermis subsp. *pumpellianus* var.
    purpurascens 135
Bromus inermis var. *aristatus* 132
Bromus inermis var. *ciliatus* 131
Bromus inopinatus 132
Bromus intermedius subsp. *divaricatus* 138
Bromus japonicus 18, **82**, 138
Bromus japonicus subsp. *anatolicus* 139
Bromus japonicus var. *porrectus* 139
Bromus japonicus var. *subsquarrosus* 139
Bromus kalmii 17, **36**, 133
Bromus kalmii var. *major* 134
Bromus kalmii var. *occidentalis* 134
Bromus kalmii var. *porteri* 134
Bromus laciniatus **130**, 131, 142
Bromus laeviglumis 135
Bromus laevipes 17, **38**, 133
Bromus lanatipes 7, 13, 15, 16, 24, **40**, 133
Bromus lanatipes f. *glaber* 16, **40**, 133
Bromus lanceolatus 18, **84**, 139
Bromus latiglumis 12, **42**, 133
Bromus latiglumis f. *incanus* **42**, 133
Bromus latior 142
Bromus lepidus 17, **85**, 139
Bromus lepidus f. *lasiolepis* 138
Bromus lloydianus 138
Bromus luzonensis 6, 104, 141
Bromus macounii 132
Bromus macrostachys 84, 139
Bromus madritensis 21, **118**, 143
Bromus madritensis var. *maximus* 143
Bromus madritensis var. *rigidus* 143
Bromus madritensis var. *rubens* 143
Bromus magnificus 134
Bromus marginatus 7, 8, 20, 94, 95, 100,
    101, 104, **106**, 110, 142
Bromus marginatus subsp. *maritimus* 142
Bromus marginatus var. *breviaristatus* 142
Bromus marginatus var. *latior* 20, **106**, 142

Bromus marginatus var. *seminudus* 20, **106**, 142
Bromus maritimus 7, 20, 94, **108**, 142
Bromus maximus 143
Bromus maximus var. *gussonei* 143
Bromus meyeri 131
Bromus molliformis 78, 130, 138
Bromus mollis 78, 137
Bromus mollis f. *leiostachys* 138
Bromus mollis var. *commutatus* 137
Bromus mollis var. *leiostachys* 138
Bromus mollis var. *racemosus* 139
Bromus mollis var. *secalinus* 139
Bromus mucroglumis 7, 13, 14, **44**, 134
Bromus mucronatus 141
Bromus multiflorus 139, 142
Bromus mutabilis var. *commutatus* 137
Bromus mutabilis var. *pratensis* 137
Bromus nanus 138
Bromus nitens 140
Bromus nottowayanus 7, 14, 16, **46**, 58, 134
Bromus orcuttianus 13, 15, **48**, 134
Bromus orcuttianus var. *grandis* 132
Bromus orcuttianus var. *hallii* 15, 16, **48**, 50, 134
Bromus oregonus 140
Bromus pacificus 13, 16, **52**, 134
Bromus paniculatus 142
Bromus parviflorus 142
Bromus patulus 138
Bromus pauciflorus 142
Bromus pendulus 139
Bromus polyanthus 7, 8, 19, 94, 95, 106, **110**, 142
Bromus polyanthus var. *paniculatus* **110**, 142
Bromus porteri 7, 8, 16, 24, **54**, 134
Bromus porteri var. *assimilis* 132
Bromus porteri var. *frondosus* 132
Bromus porteri var. *havardii* 131
Bromus porteri var. *lanatipes* 133
Bromus pratensis 137
Bromus pratensis var. *apricorum* 137
Bromus pseudolaevipes 7, 17, **56**, 135
Bromus x pseudothominei 138
Bromus pubescens 13, 14, **58**, 135
Bromus pumpellianus 7, 8, 12, **60**, 135
Bromus pumpellianus subsp. *dicksonii* **60**, 61, 136
Bromus pumpellianus var. *arcticum* **60**, 136

Bromus pumpellianus var. *melicoides* 136
Bromus pumpellianus var. *tweedyi* 135
Bromus pumpellianus var. *villosissimus* **60**, 136
Bromus purgans 58, 133, 135
Bromus purgans f. *glabriflorus* 135
Bromus purgans f. *laevivaginatus* 135
Bromus purgans var. *incanus* 133
Bromus purgans var. *laeviglumis* 135
Bromus purgans var. *latiglumis* 133
Bromus purgans var. *longispicatus* 135
Bromus purgans var. *pallidus* 131
Bromus purgans var. *purpurascens* 135
Bromus purgans var. *texensis* 136
Bromus purgans var. *vulgaris* 136
Bromus racemosus 18, **86**, 139
Bromus racemosus subsp. *commutatus* 137
Bromus racemosus var. *commutatus* 137
Bromus ramosus 130
Bromus richardsonii 7, 8, 14, 16, 26, **62**, 136
Bromus richardsonii var. *pallidus* 131
Bromus rigidus 8, 20, 116, **120**, 143
Bromus rigidus var. *gussonei* 116, 143
Bromus riparius 8, 28
Bromus rubens 21, **122**, 143
Bromus rubens var. *rigidus* 143
Bromus scabratus 134
Bromus schraderi 141
Bromus secalinus 8, 17, **90**, 139
Bromus secalinus var. *gladewitzii* 137
Bromus secalinus var. *velutinus* **90**, 140
Bromus setaceus 144
Bromus scoparius 18, **88**, 139
Bromus scoparius var. *rubens* 143
Bromus sitchensis 19, 94, 95, **112**, 142
Bromus sitchensis var. *aleutensis* 140
Bromus sitchensis var. *marginatus* 142
Bromus squarrosus 18, **92**, 140
Bromus squarrosus var. *patulus* 139
Bromus squarrosus var. *racemosus* 139
Bromus stamineus 19, 94, **114**, 143
Bromus sterilis 20, **124**, 143
Bromus steudelii 133
Bromus strictus 141
Bromus submuticus 139
Bromus subvelutinus 6, 7, 19, 94, 95, 100, **104**, 106, 141
Bromus suksdorfii 12, **64**, 136
Bromus tectorum 8, 20, **126**, 144

Bromus tectorum f. *nudus* **126**, 144
Bromus tectorum var. *glabratus* 144
Bromus tectorum var. *nudus* 144
Bromus texensis 12, **66**, 136
Bromus thominei 138
Bromus trinii 128, 144
Bromus trinii var. *excelsus* 128, 144
Bromus trinii var. *pallidiflorus* 144
Bromus unilateralis 139
Bromus unioloides 141
Bromus unioloides var. *haenkeanus* 141
Bromus velutinus 140
Bromus vestitus 138
Bromus villosus 143
Bromus villosus var. *gussonei* 143
Bromus villosus var. *maximus* 143
Bromus villosus var. *rigidus* 143
Bromus virens 140
Bromus virens var. *minor* 141
Bromus vulgaris 13, 15, **68**, 130, 136
Bromus vulgaris var. *eximius* **68**, 136
Bromus vulgaris var. *robustus* **68**, 136
Bromus willdenowii 141
Bromus wolgensis 140
Ceratochloa arizonica 140
Ceratochloa breviaristata 141
Ceratochloa carinata 140
Ceratochloa cathartica 141
Ceratochloa grandiflora 140
Ceratochloa haenkeana 141
Ceratochloa laciniata 142
Ceratochloa marginata 142
Ceratochloa maritima 142
Ceratochloa pendula 141
Ceratochloa submutica 141
Ceratochloa unioloides 141
Danthonia pseudo-spicata 144
Festuca erecta 132
Festuca inermis 132
Festuca inermis var. *villosa* 132
Festuca madritensis 143
Festuca rubens 143
Festuca unioloides 141
Forasaccus arvensis 137
Forasaccus brebiaristatus 141
Forasaccus ciliatus 131
Forasaccus ciliatus var. *laeviglumis* 135
Forasaccus commutatus 137
Forasaccus erectus 132
Forasaccus inermis 132

Forasaccus latiglumis 133
Forasaccus marginatus 142
Forasaccus maximus 143
Forasaccus mollis 137
Forasaccus patulus 139
Forasaccus pumpellianus 135
Forasaccus purgans 133
Forasaccus racemosus 139
Forasaccus secalinus 139
Schedonorus inermis 132
Schedonorus sterilis 143
Schedonorus tectorum 144
Schedonorus unioloides 141
Serrafalcus arvensis 137
Serrafalcus commutatus 137
Serrafalcus lloydianus 138
Serrafalcus macrostachys 139
Serrafalcus molliformis 138
Serrafalcus mollis 137
Serrafalcus neglectus 137
Serrafalcus patulus 138
Serrafalcus racemosus 139
Serrafalcus racemosus var. *commutatus* 137
Serrafalcus scoparius 139
Serrafalcus secalinus 139
Tragus unioloides 141
Trisetobromus hirtus 144
Trisetum barbatum 144
Trisetum barbatum var. *major* 144
Trisetum hirtum 144
Trisetum trinii 144
Trisetum trinii var. *majus* 144
Trisetum trinii var. *pallidiflorus* 144
Zerna anomala 131
Zerna ciliata 131
Zerna erecta 132
Zerna gussonei 143
Zerna inermis 132
Zerna latiglumis 133
Zerna macrostachys 139
Zerna madritensis 143
Zerna pumpelliana 135
Zerna purgans 133
Zerna richardsonii 136
Zerna rubens 143
Zerna sterilis 144
Zerna tectorum 144
Zerna unioloides 141
Zerna vulgaris 136